Inorganic Spectroscopic Methods

Alan K. Brisdon

Department of Chemistry, UMIST, Manchester

Series sponsor: **ZENECA**

ZENECA is a major international company active in four main areas of business: Pharmaceuticals, Agrochemicals and Seeds, Speciality Chemicals, and Biological Products.

ZENECA's skill and innovative ideas in organic chemistry and bioscience create products and services which improve the world's health, nutrition, environment, and quality of life.

ZENECA is committed to the support of education in chemistry and chemical engineering.

OXFORD NEW YORK TOKYO
OXFORD UNIVERSITY PRESS
1998

Oxford University Press, Great Clarendon Street, Oxford OX2 6DP

Oxford New York
Athens Auckland Bangkok Bogota Bombay Buenos Aires
Calcutta Cape Town Dar es Salaam Delhi Florence Hong Kong Istanbul
Karachi Kuala Lumpur Madras Madrid Melbourne Mexico City
Nairobi Paris Singapore Taipei Tokyo Toronto Warsaw

and associated companies in
Berlin Ibadan

Oxford is a trade mark of Oxford University Press

Published in the United States
by Oxford University Press Inc., New York

A catalogue record for this book is available from the British Library

Library of Congress Cataloging in Publication Data
(Data available)

ISBN 0 19 855949 6

Typeset by the author

Printed in Great Britain by
The Bath Press, Bath

Series Editor's Foreword

In the process of establishing the structure and properties of an inorganic compound, an array of spectroscopic techniques is required. Due to the varied nature of inorganic compounds, the approaches are often more diverse than that utilized within organic chemistry; the link to symmetry is generally stronger, both for vibrational and magnetic resonance spectroscopies. The accent of this book is on showing how these techniques are utilized in inorganic chemistry.

Oxford Chemistry Primers are designed to give a concise introduction to all chemistry students by providing the material that would usually be covered in a 8–10 week lecture course. As well as giving up-to-date information, this series provides explanations and rationales that form the framework of understanding inorganic chemistry. Alan Brisdon here provides a general handbook for spectroscopic characterization of inorganic compounds. It provides an accessible and thorough guide to the application of spectroscopic techniques that should be of great assistance to chemistry students right through their undergraduate career.

John Evans
Department of Chemistry
University of Southampton

Preface

An understanding and ability to interpret spectroscopic data is a pre-requisite for students in chemistry and related disciplines from the undergraduate level onwards. The aim of this book, therefore, is to provide a working knowledge of the common spectroscopic techniques and their application to inorganic-based systems. In a book of this size, and at this level, it is not possible to demonstrate the huge range of spectroscopic applications across the whole of inorganic chemistry but merely an introduction to the areas most commonly encountered in undergraduate courses. The approach taken is unashamedly aimed at the application of the techniques and interpretation of the spectra obtained rather than providing a theoretical treatment of the methods; these can be found in the textbooks referenced at the end of each chapter.

The starting point is a description of electromagnetic radiation and its interaction with matter as this is a common feature for many of the techniques discussed later. Each subsequent chapter covers related spectroscopic methods and includes a brief, and non-rigorous, look at their physical basis and applications typical in inorganic compounds. A number of worked examples are given in the text along with short questions in the margin to test the basics. Because it is rarely wise to rely on any single technique to completely identify unknown materials, the last chapter looks at problems which require the application of a number of different spectroscopic methods to provide a complete solution.

I would like to thank all my colleagues, staff, and students who have read, commented and corrected parts of the text, provided spectra and ideas. Particular thanks go to the postgraduate and undergraduate students who have worked in my research group. Finally I want to thank my wife for her patience and understanding while I have been writing this book.

Manchester A.K.B.
October 1997

Contents

1 Introduction

1.1 Why spectroscopy?

Chemists, amongst other things, synthesise materials; but having prepared something, it needs to be identified. Elemental analysis can provide the ratios of the elements in the compound (or mixture) but tells us nothing more. The most direct method of obtaining the structure of a pure compound in the solid phase is single-crystal X-ray structure determination. This can provide a complete picture of the compound of interest—but there may be problems. What if suitable crystals cannot be grown? What if the material to be studied is, under normal conditions, a solution, a liquid, or a gas? Even if we do obtain crystals can we be sure that they are representative of the solution they came from, or is it just the least soluble of a series of products precipitating out? There is, however, a more important factor—time. The whole process of growing crystals, obtaining and analysing their diffraction patterns is a time-consuming business.

Usually we do not need the full structure of every reaction product or intermediate, we may only need to determine whether the reaction is complete, if a particular group or ligand is present in the molecule, whether a product is pure, or if the reaction conditions have attacked one part of the molecule rather than another. It is to answer these questions that we turn to one or more of the available spectroscopic techniques.

The piecing together of information provided by spectroscopic methods to provide the identity and structure of a compound has frequently been compared with attempting to solve a cryptic crossword puzzle. This comparison may, in part, be correct; in both cases we try to solve a problem based on indirect information.
Most spectroscopic problems can be solved with an understanding of the processes involved, some background information, and a logical approach.
There is certainly one common feature to solving crossword and spectroscopic problems—the more practice you get the better you become!

1.2 Radiation and matter

A dictionary definition of spectroscopy might be that it is the study of the interaction between radiation and matter. Since the techniques described in this book, with the exception of mass spectrometry, involve such an interaction we start by considering electromagnetic (em) radiation and some of its properties.

Such radiation may be represented as two oscillating components, one electric and, at 90°, an associated magnetic component. These propagate in the direction perpendicular to both of these waves as represented in Fig. 1.1. Since our definition of spectroscopy is the interaction of matter with radiation, then to observe a spectrum the compound must, as a minimum requirement, interact either with the electric or magnetic component of the applied radiation.

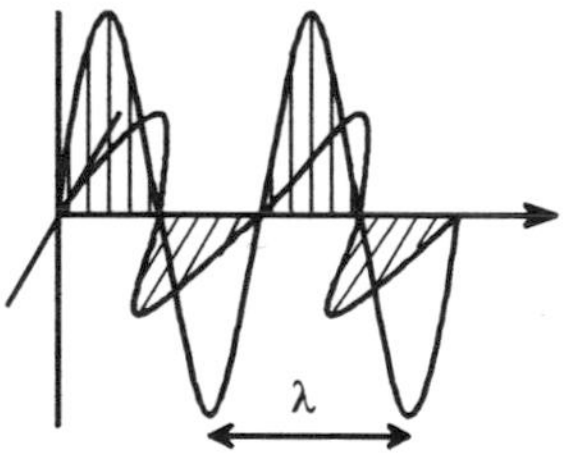

Fig. 1.1 A representation of electromagnetic radiation.

Wave-like electromagnetic radiation may be defined in terms of either its characteristic frequency, ν, or wavelength, λ, which are related by

$$\lambda \times \nu = c \quad (1.1)$$

where c is the speed of light, 2.998×10^8 m s^{-1}.

The *electromagnetic spectrum* covers a very wide range of energies from X-rays possessing short wavelengths (and high frequencies) through visible light to longer wavelength radio frequencies. A representation of the span of energies, frequencies, and related wavelengths is shown in Fig. 1.2.

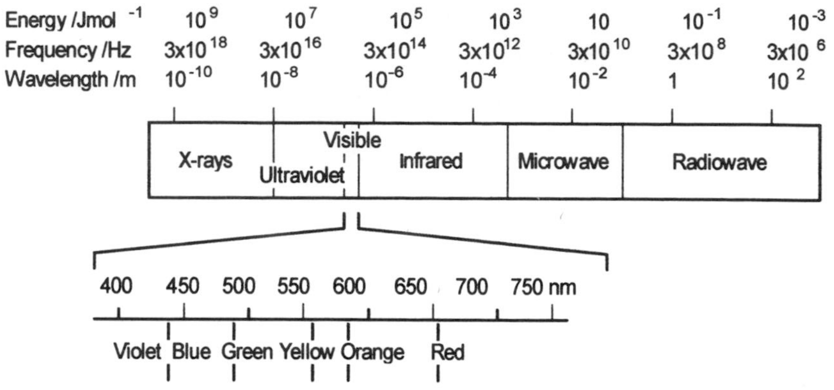

Fig. 1.2 The electromagnetic spectrum.

Under certain circumstances it is better to consider radiation as a series of 'packets of energy' or *photons* of radiation, rather than as a continuous wave; this is the so-called wave–particle duality of radiation proposed by de Broglie. The energy of each photon is given by

$$E = h\nu \tag{1.2}$$

where h is Planck's constant, 6.626×10^{-34} J s, and ν the frequency of the radiation.

Thus it is possible using eqns 1.1 and 1.2 to convert between frequency and wavelength and to calculate the energy of each photon of light of a particular frequency.

For example, the frequency of visible light with a wavelength of 400 nm is

$$\begin{aligned}\nu &= c / \lambda \\ &= 3 \times 10^{8}\ \text{m s}^{-1} / 400 \times 10^{-9}\ \text{m} \\ &= 7.5 \times 10^{14}\ \text{s}^{-1}\end{aligned}$$

The energy of each light photon is

$$\begin{aligned}E &= h\nu \\ &= 6.626 \times 10^{-34}\ \text{J s} \times 7.5 \times 10^{14}\ \text{s}^{-1} \\ &= 4.970 \times 10^{-19}\ \text{J}\end{aligned}$$

When this value is multiplied by the Avogadro constant, the energy for a mole of such photons is obtained

$$\begin{aligned}&4.97 \times 10^{-19}\ \text{J} \times 6.022 \times 10^{23}\ \text{mol}^{-1} \\ &= 299.3\ \text{kJ mol}^{-1}\end{aligned}$$

An alternative unit often used in spectroscopic work instead of frequency or wavelength is the *wavenumber*; this is the number of waves that would occupy a 1 cm length and therefore has the dimension of cm^{-1} and is usually denoted by the symbol $\bar{\nu}$. For example, light of wavelength 400 nm can be expressed in terms of wavenumbers by first converting to a wavelength in centimetres and then taking the reciprocal.

$$\begin{aligned}\lambda &= 400 \text{ nm} \\ &= 400 \times 10^{-9}\text{ m} \\ &= 400 \times 10^{-7}\text{ cm} \\ \bar{\nu} &= 1/(400 \times 10^{-7}\text{ cm}) = 25{,}000\text{ cm}^{-1}\end{aligned}$$

Question 1.1 What is the wavelength (in m) and energy (in kJ mol^{-1}) associated with the magnetic resonance of a proton at 100 MHz?

Although the wavenumber is not an SI unit, its use persists on two counts. Firstly, it makes for more convenient numbers, especially in the infrared section of the electromagnetic spectrum (where its use is most widespread) and, secondly, because the frequency of radiation expressed in wavenumbers is directly related to energy.

Question 1.2 What is the wavelength of radiation of $\bar{\nu}$ = 2000 cm^{-1} ?

Molecular energy levels

We now turn our attention to matter; chemists are concerned with compounds which are composed of a number of atoms which in turn contain a number of subatomic particles, for example electrons and protons. Because molecules and their constituent parts are so small, the classical rules of physics in which bodies may possess any energy do not apply, instead only certain discrete energy levels are possible. These levels and their rules are described by quantum mechanics and are identified by a set of unique *quantum numbers*. So just as there is a set of quantum numbers (N, l, m_l and m_s) that describe the type of orbital with which an electron is associated, there are quantum numbers that describe, for example, in which vibrational or rotational state a molecule exists, Table 1.1.

Quantum number	Symbol	Values
Principal	N	1,2,3,...
Azimuthal	l	0...N–1
Magnetic	m_l	l...0...–l
Electron spin	m_s	+½, –½
Vibrational	v	0,1,...
Rotational	J	0,1,...
Nuclear spin	I	n/2 (n=1,2,...)

Table 1.1 Some of the possible quantum numbers used to describe electrons and molecules.

Because the energy levels associated with a molecule are discrete, transitions between the various energy states will only occur at certain discrete energies. In theory, a transition between two different energy levels can be caused by subjecting the molecule to radiation with energy that *exactly* matches the difference between the two energy levels, as given by

$$E_{\text{upper}} - E_{\text{lower}} = \Delta E = h\nu \tag{1.3}$$

There are two possible forms such transitions may take, as shown in Fig. 1.3. If radiation of the correct frequency is supplied, then a transition occurs from a lower state (usually the lowest occupied, or *ground state*) to a higher energy state, Fig. 1.3(a). In this case energy is absorbed by the sample giving rise to an *absorption* process. Conversely, we could cause a higher energy, or *excited*, state to be occupied (for example by heating the sample or providing a short pulse of energy) and then record the energy emitted as the molecule returns, or *relaxes*, to the ground state which would result in the *emission* of energy from the sample, Fig. 1.3(b). It is the study of the

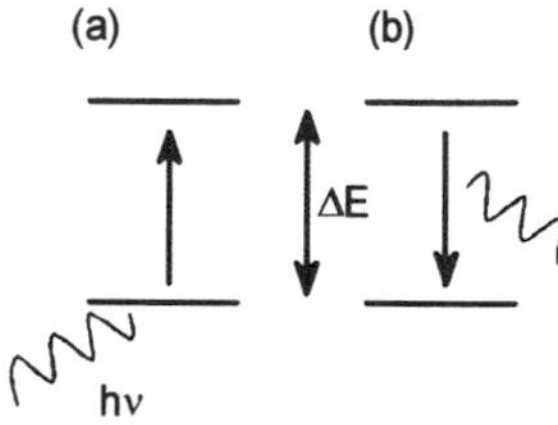

Fig. 1.3 Processes occurring in (a) absorption and (b) emission spectroscopy.

absorption or emission of energy arising from these changes in energy states that is the basis of all true spectroscopic techniques.

There is a very wide range of possible transitions that may be investigated, for example we could study processes that affect the whole of the molecule, such as a change in shape or orientation, or transitions of some of the variety of much smaller components such as electrons or nuclei. Because these particles, atoms, and molecules have very different masses, transitions occur at very different energies and according to the Born–Oppenheimer approximation they can be considered as separate.

Figure 1.4 shows a representation of some of the differing energy levels in molecules. Because of the differences in the magnitude of the energies involved, the energy of a molecule can be separated, or *partitioned*, into the sum of individual contributions, for example

$$E = E_{\text{electronic}} + E_{\text{vibration}} + E_{\text{rotation}} + E_{\text{translation}}$$

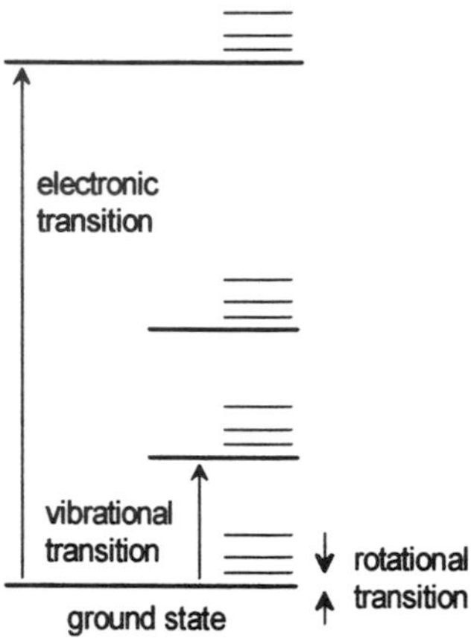

Fig. 1.4 A sketch of the separation of various energy levels for a molecule.

The energies that induce electronic transitions range from γ-rays to the ultraviolet; changes in vibrational states occur at lower energies, typically within the infrared part of the electromagnetic spectrum, and at still lower energies rotational transitions occur.

Table 1.2 lists a number of the more common spectroscopic techniques and the underlying physical processes that occur during the experiment. The range of energies required for such transitions is given, as is the wavelength of the radiation involved.

Wavelength range (m)	Energy range (J mol^{-1})	Time-scale (s^{-1})	Spectroscopic technique	Processes occurring
(2–9) × 10^{-7}	(1–6) × 10^{5}	10^{-14}	UV–visible	Change in electronic distribution in metal ions, in ligands and between them
(2–50) × 10^{-6}	(2–60) × 10^{3}	10^{-13}	Infrared (IR)	Vibrations of molecules and groups within molecules
(6–600) × 10^{-5}	20–2000	10^{-11}	Rotational (Microwave)	Rotation of complete molecules around molecular axes
(9–30) × 10^{-3}	4–15	10^{-4}–10^{-8}	Electron spin resonance (ESR)	Change in spin of unpaired electrons in molecules and radical ions
0.6–5	(2–20) × 10^{-2}	10^{-1}–10^{-8}	Nuclear magnetic resonance (NMR)	Change in spin of magnetically active nuclei (I > 1/2)
5–15	(8–20) × 10^{-3}	10^{-3}–10^{-9}	Nuclear quadrupole resonance (NQR)	Change in quadrupole orientation

Table 1.2 Common spectroscopic techniques, their uses and energy ranges.

Ideally, a spectroscopic transition will occur between molecules in a single well-defined ground state to a single higher-energy state corresponding to an increase in just one quantum number. Frequently, however, this is not the case. This may be because there is more than one significantly populated lower energy state, as described in the next section, or because it is possible for excitation in more than one state to occur. So, for example, electronic transitions may be accompanied by changes in vibrational energies and rotational transitions may accompany vibrational absorptions. Usually when this occurs a number of peaks are observed (although they may not be clearly separated). We therefore see many absorptions, referred to as *fine structure*, in a narrow energy range, as shown in Fig. 1.5.

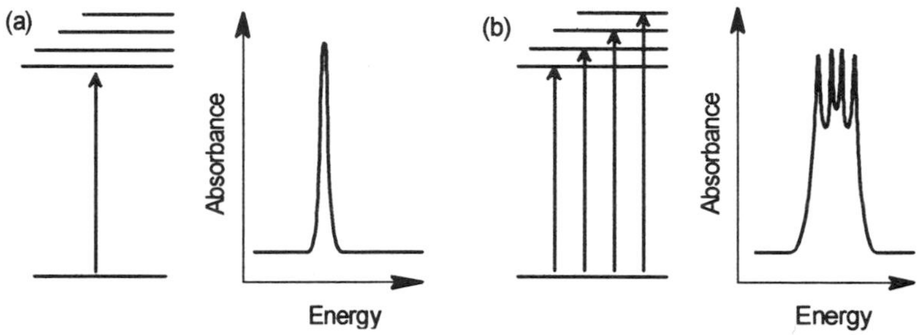

Fig. 1.5 Transitions which result in changes in two, or more, different quantum numbers give rise to fine structure on peaks in spectra.

Question 1.3 What type of fine structure might you expect to observe on an ESR spectrum?

Population distributions

Consideration of the population distribution of a set of energy levels is important from a spectroscopic point of view because of the need to know which levels are significantly occupied and hence likely to undergo absorption or emission of energy. If the upper and lower energy levels are equally populated then absorption and emission are both equally likely and so no spectroscopic transition will be seen. In order to record a spectrum there has to be a difference in the population of the two states involved. In a similar way a liquid is only observed to flow between two connected containers when there is a difference in their levels, Fig. 1.6.

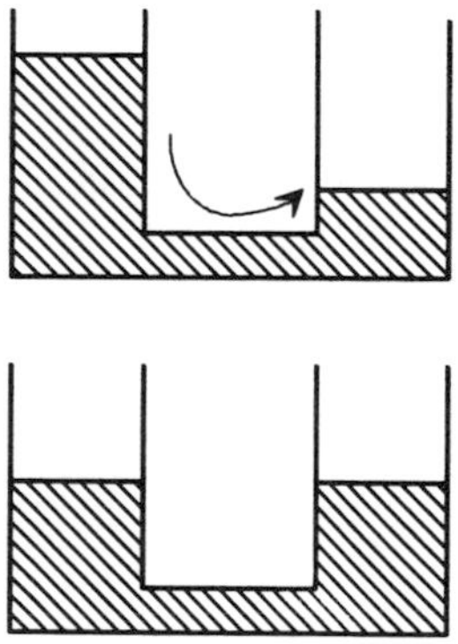

Fig. 1.6 Spectroscopic transitions only occur when there is a difference in populations, just as water only flows when the two levels differ.

On the microscopic scale the ratio of the equilibrium population in two states separated by an energy gap of ΔE is given by the Boltzmann distribution

$$N_{\text{upper}}/N_{\text{lower}} = \exp(-\Delta E/kT) \qquad (1.4)$$

where k is the Boltzmann constant (1.381×10^{-23} J K^{-1}) and T the temperature of observation expressed in Kelvin.

For example, in infrared vibrational spectroscopy the difference between two energy states for a band that absorbs at 2000 cm^{-1} is around 4×10^{-20} J, so at room temperature the population distribution will be

$$\begin{aligned} N_{\text{upper}}/N_{\text{lower}} &= \exp\left(-4\times10^{-20}\ \text{J}/1.381\times10^{-23}\ \text{J K}^{-1}\times 298\ \text{K}\right) \\ &= \exp(-9.722) \\ &= 6.0\times10^{-5} \end{aligned}$$

This means that nearly all the molecules exist in the lower vibrational state. By contrast, for two states separated by an energy gap of 1.33×10^{-25} J (such as that for two different NMR spin states) the population ratio is 0.999975, or an almost equal occupation of the upper and lower energy levels. On this basis we would expect vibrational spectroscopy to be more sensitive than NMR spectroscopy simply because there is a greater imbalance of the occupied energy states and hence a greater probability of a transition occurring.

Question 1.4 What is the population ratio for two electronic states separated by 450 nm at 298 K?

Selection rules

Besides the requirement that there is a population difference between ground and excited states for a transition to occur, there is another important factor. This arises from the use of quantum theory to describe the properties of small particles which in turn imposes certain limitations on which transitions should occur. The rules that govern which of these transitions have a finite (i.e. non-zero) probability of occurring are called *selection rules.* Although these rules vary from technique to technique, in general transitions between two different energy states in which a single quantum number, such as v, the vibrational quantum number, has altered only by ± 1 are usually allowed. So a transition between two different particular vibrational states $v = 1 \rightarrow v = 2$ will be formally *allowed* by the selection rule for vibrational spectroscopy that $\Delta v = \pm 1$ and the transition from the ground state to the first excited state (e.g. $v = 0 \rightarrow v = 1$) is called the *fundamental* transition.

These selection rules are a formal statement of which transitions are expected (*allowed*) or not expected (*forbidden*) to be observed. However, in many cases there are perturbations to the ground or excited states or alternative mechanisms which result in formally forbidden transitions being observed, albeit at lower intensity than those arising from allowed events. For example, transitions from the ground state to the $v = 2$ or $v = 3$ levels are called *overtones* and are usually only weak bands.

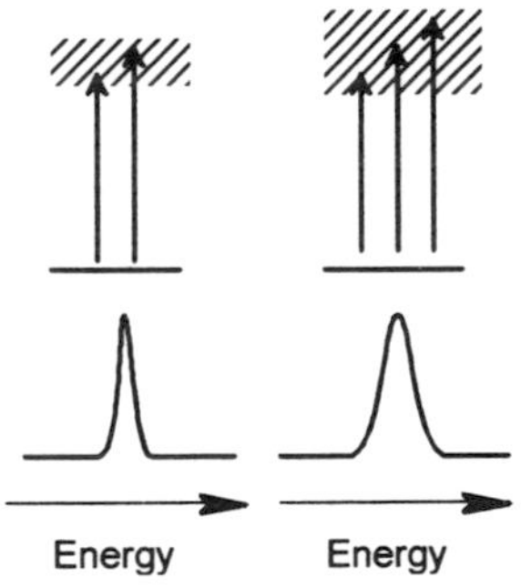

Fig. 1.7 The uncertainty of the energy level associated with a short relaxation time results in broad peaks in spectra.

Time-scales

Time is important as far as spectroscopic techniques are concerned in a number of respects. A certain time is required for the interaction of a photon of energy with the molecule to cause excitation. For all the processes we will consider this is very fast, in the order of 10^{-18} s, which is much quicker than any electronic transition, molecular rearrangement or vibration.

Once excitation has occurred, resulting in a higher energy state, the excess energy is lost over a certain period of time to revert to the ground state. This is known as the *relaxation time* and varies from one technique to another, and some typical values are given in Table 1.2. If the energy is lost before the molecules have a chance of changing (by vibrating or altering shape, etc.), then the spectroscopic method 'sees' molecules in a number of different states. However, if the relaxation rate is slow then molecules can undergo their normal motion and the technique will effectively see a single 'average' set of molecules. NMR spectroscopy is one method where long relaxation times can have a profound effect on the observed spectra and this is covered in more detail in Chapter 3.

The effects of short relaxation times is more widespread and arises from the *Uncertainty Principle*. This limits the precision with which both the lifetime of the excited state and its relative energy can be determined, as given by eqn 1.5,

$$\tau\,\Delta E \leq h/2\pi \tag{1.5}$$

Thus as the lifetime (τ) of the excited state becomes shorter, the uncertainty in the accuracy of the energy level difference (ΔE) becomes greater. The range of energies absorbed therefore becomes larger and the peak in the spectrum corresponding to this transition becomes wider, as shown in Fig. 1.7. There are a number of factors which may contribute to the actual lifetime of an excited state, such as the mechanism of relaxation, which itself will be influenced by the physical state of the sample. These factors are dealt with in more detail, where appropriate, in later chapters.

Question 1.5 Which of the spectroscopic methods in Table 1.2 will possess the broadest bands due to uncertainty?

1.3 Instrumentation

The basic scanning spectrometer may be represented pictorially as shown in Fig. 1.8.

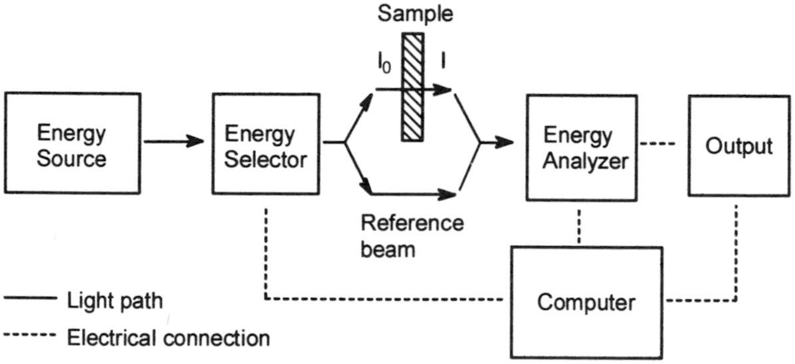

Fig. 1.8 The basic components of a scanning spectrometer.

This consists of either a tuneable source of suitable energy or a method of selecting energies produced by a wide-range source. The monochromatic energy is frequently split into two paths with one passing through the sample and the other acting as a reference beam. Energy will be absorbed from the sample beam when it exactly matches a transition of the sample. The remaining electromagnetic energy then reaches a detector so that the amount of energy absorbed can be measured by comparing it with the reference beam and this is presented on the chart recorder or computer screen. This is the basis of *absorption spectroscopy* and is commonly encountered in vibrational and electronic spectroscopy.

In the sort of experiment described above, the final spectrum will usually be a plot of absorption (A), a measure of how much energy has been absorbed, versus energy. Alternatively, the proportion of energy that has passed through the sample, the percentage transmission (T), can be plotted against energy. Illustrations of these types of spectra are shown in Fig. 1.9.

There is an important distinction between these two ordinate (y) axis scales. The mathematical functions that give rise to them are shown below, where I_0 is the intensity of the applied energy and I the resultant intensity after passing through the sample.

$$A = \log_{10}\frac{I_0}{I} \qquad T = \frac{I_0}{I} \times 100$$

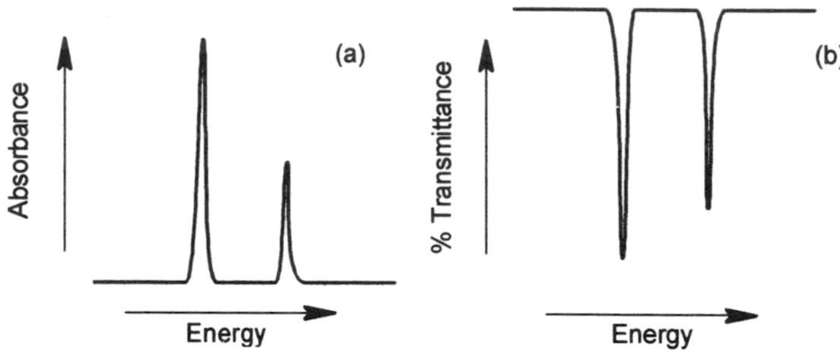

Fig. 1.9 Typical (a) absorption and (b) transmission spectra for the same sample.

If a sample of a particular compound absorbs 20% of the energy at a certain frequency then 80% will pass through to the detector. If a second identical sample is additionally placed in the sample beam then a further 20% of the remaining energy, i.e. 16% (20% of 80%), would be absorbed and so 64% (100% – 20% – 16%) of the initial energy will be transmitted. Thus, doubling the amount of absorbing compound results in less than a doubling of the observed peak intensity on the percentage transmission scale. However, a doubling of the sample concentration results in a doubling of the absorption and this can be confirmed if we convert from percentage transmission to absorption values using the above equation.

$$A_1 = \log_{10}\left(\frac{100\%}{80\%}\right) = 0.097 \qquad A_2 = \log_{10}\left(\frac{100\%}{64\%}\right) = 0.194$$

For this reason most spectroscopic techniques present the ordinate axis of spectra in terms of absorption, rather than transmission, the common exception being IR spectroscopy. It is essential if relative concentrations or compositions are to be deduced, or that spectral subtractions are to be performed, that spectra are compared in absorbance or related units which have a linear relationship with concentration.

Fourier transforms

One of the disadvantages of the scanning technique described above is that it is relatively slow, especially for high-resolution studies. In order that a complete spectrum is recorded, all energies within the spectroscopic range are successively generated or selected from a source and directed towards the sample in sequence and the absorption of the energy measured. It is more common these days, especially for vibrational and NMR studies, for an alternative method to be used which essentially allows all the data for the complete spectrum to be recorded very rapidly. Such techniques use *Fourier transform* methods to mathematically de-convolute individual absorptions at particular energies (or frequencies) from data containing all such energies which were recorded as an interference pattern.

These mathematical operations, described by the French mathematician Fourier, provide a method of changing from one domain to another. For example, a sine wave represented as a function of time (s) is shown in Fig. 1.10(a), Fourier transformation of this sine wave converts it to the reciprocal domain (s^{-1} or Hz) shown in Fig. 1.10(b) which is the frequency of the waveform. In spectroscopic systems the Fourier transform is used to convert data arising from superposition of many different waves, Fig 1.11(a), to their characteristic frequencies, Fig. 1.11(b), simultaneously.

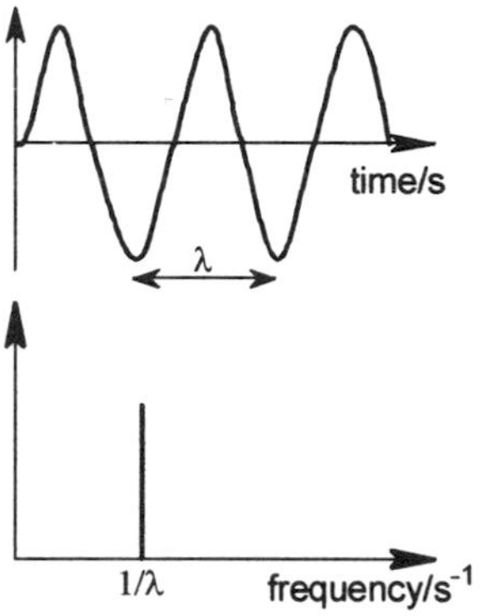

Fig. 1.10 (a) A sine wave and (b) its Fourier transform.

There are a number of advantages of using Fourier transform techniques in spectroscopy and the principle one is that the whole of the spectrum is recorded in one go. So for the time it takes to scan through the complete spectrum sequentially, a number of interferograms can be recorded and added together to improve the signal-to-noise ratio. As a series of scans are added the random noise is reduced in intensity and the signals due to peaks are reinforced. In this way the more scans recorded the better the signal-to-noise ratio (S/N) in the final spectrum, such that after n scans the S/N has increased by $\sqrt{n}$.

Most modern spectrometers, be they of the Fourier transform type or otherwise, are connected to computers so that data may be stored and manipulated electronically. This facilitates smoothing noisy data, removing background peaks, and making the measurement of spectroscopic parameters, such as peak positions and relative intensities, much easier. It also makes possible the use of computer libraries for comparison and identification purposes.

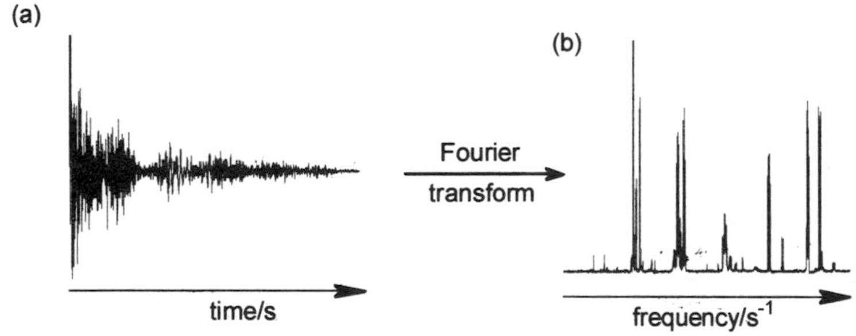

Fig. 1.11 (a) A superposition of waveforms typically observed in NMR spectroscopy and (b) the Fourier transform.

Further reading

W.G. Richards and P.R. Scott, Energy levels in atoms and molecules (Oxford Chemistry Primer Series), Oxford University Press (1994).

C.N. Banwell, Fundamentals of molecular spectroscopy, McGraw-Hill, London (1972).

E.A.V. Ebsworth, D.W.H. Rankin and S. Craddock, Structural methods in inorganic chemistry, Blackwell Scientific Publishers (1986).

2 Vibrational spectroscopy

2.1 The basics

Vibrational spectroscopy was one of the first spectroscopic techniques to be widespread in its use, in particular infrared or IR spectroscopy—so called because of the region of the electromagnetic spectrum used. There is a second form of vibrational spectroscopy—Raman spectroscopy—which provides similar information but has a different physical basis and is covered in more detail later in this chapter. As the name suggests these forms of spectroscopy are tied up with changes in the *vibrational* state of molecules.

In a simple diatomic molecule, such as carbon monoxide, shown in Fig. 2.1, the two atoms are held together by the overlap of a number of orbitals. At a certain internuclear distance there is a balance between the attractive bonding forces and the repulsive interaction between the remaining electrons of the two atoms. This equilibrium bond distance (usually denoted as r_e, where e refers to the equilibrium position) can be changed by applying energy. In this respect we could think of this molecule as being two balls (of mass m_1 and m_2) held together by a spring of a certain strength (k). Classical physics tells us that such a system will vibrate at a particular frequency, ω, given by eqn 2.1

Fig. 2.1 The structure of carbon monoxide.

$$\omega = \frac{1}{2\pi}\sqrt{\frac{k}{\mu}} \tag{2.1}$$

where μ is the reduced mass of the two objects, given by eqn 2.2

$$\mu = \frac{(m_1 \times m_2)}{(m_1 + m_2)} \tag{2.2}$$

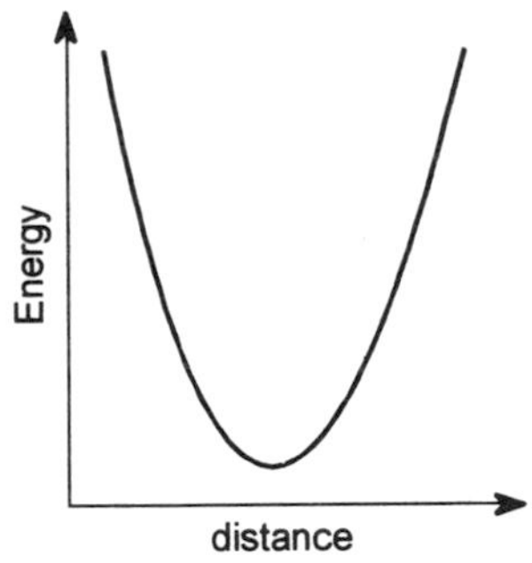

Fig. 2.2 A potential well for a classical vibrator.

This macroscopic view of an oscillator would allow us to sketch an energy diagram similar to that shown in Fig. 2.2. Although this is a useful analogy it is limited in a number of respects. Firstly, on the atomic scale quantum theory requires that only certain energy levels may exist, in other words our 'molecular-spring' can only be stretched in steps of a certain size, given by eqn 2.3. In fact, the lowest energy state, v=0, does not possess zero vibrational energy; if it did it would contradict the uncertainty principle. Instead the energy is given by

$$E = h\omega\,(v + \tfrac{1}{2}) \tag{2.3}$$

where v is the vibrational quantum number.

Repulsion between the two nuclei and their electrons makes it much more difficult to bring the two atoms very close to each other than the simple model predicts. Conversely, if we put sufficient vibrational energy into the molecule the atoms will vibrate so much that eventually the atoms will become far enough apart that the attractive forces are too weak to keep them together and so the bond *dissociates*. Figure 2.3 shows a modified, Morse, potential-well diagram that takes these factors into account and is more realistic for diatomic molecules. As can be seen by comparing Figs 2.2 and 2.3, provided we do not consider the extremes, the behaviour is similar and so we can still use eqn 2.1 as a guide to the factors involved in the frequency at which pairs of atoms will vibrate. In particular, this tells us that the vibrational frequency will depend on the mass of the two atoms at either end of the bond and the strength of the bond, or *force constant*, holding them together.

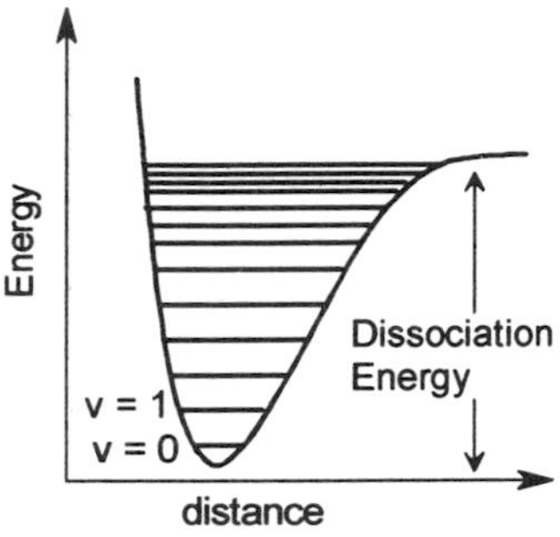

Fig. 2.3 A modified potential well for a diatomic.

We should expect heavy atoms held by weak bonds to vibrate at lower frequencies than lighter atoms and those held by multiple bonds. On this basis we would predict that the vibrational frequency of HCl will be higher than that of the deuterated analogue, DCl, which is indeed so [ν(HCl) = 2990 cm^{-1} and ν(DCl) = 2145 cm^{-1}]. Table 2.1 shows how the N–O vibrational frequency varies as the charge on the nitrogen monoxide molecule alters. As the number of electrons in the molecule alters, the occupation of the bonding and non-bonding molecular orbitals varies, so altering the N–O bond order. As the bond order and hence the bond strength increase, so does the observed vibrational frequency.

Question 2.1 Which of the following pairs of molecules would you expect to have the highest vibrational frequency: (i) BrCl, ICl; (ii) Br_2, Br_2^+; (iii) NH_3, PH_3?

Molecule	ν (cm^{-1})	Bond order
NO^+	2273	3
NO	1880	2.5
NO^-	1365	2
NO^{2-}	886	1.5

Table 2.1 Vibrational frequency of the nitrogen monoxide molecule and related ions.

The number and type of vibrations for a molecule

The position of a single atom in space can be defined with respect to an arbitrary origin by defining its distance along three axes, traditionally labelled *x*, *y*, *z*. A group of *N* atoms would require 3*N* such distances, or coordinates, to define their relative positions. We can distort the molecule by moving any atom in any of the three cartesian directions and thus a molecule of *N* atoms is described as possessing 3*N* *degrees of freedom*. For the study of vibrational spectroscopy we are only interested in the movement of the atoms such that the molecule is distorted either by altering the bonds or angles within the molecule. Not all the possible movements of the atoms will result in such a distortion, for example if every atom in the molecule moves by a similar amount in the *x*, *y*, or *z* direction then the whole molecule is merely translated. Similarly, if the atoms are moved so that the complete molecule is rotated around the *x*, *y*, or *z* axes then again this will not correspond to a change in bond lengths or angles within the molecule. So neither translation nor rotation results in a change in the shape of the molecule and, therefore, of the original 3*N* degrees of freedom, three are taken up by translations and three by rotations leaving 3*N*–6 for vibrations. Thus water, being composed of three atoms, will possess a total of nine degrees of freedom of which three [3×3–6] correspond to vibrations.

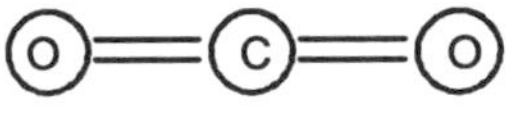

Fig. 2.4 The structure of carbon dioxide.

There is, however, an important proviso if a molecule is linear, such as carbon dioxide which is shown in Fig. 2.4. In these cases there are only two rotational degrees of freedom, since rotating the molecule around the O–C–

O axis by any amount does not appear to rotate the molecule at all, so $3N-5$ vibrational modes are expected.

Question 2.2 What is the total number of vibrational modes expected for: (i) H_2S; (ii) CO; (iii) NH_3; (iv) $Ni(CO)_4$?

Vibrations arise from changes in bond lengths and angles and in many simple, non-cyclic molecules these vibrational modes can be thought of separately and are referred to as *bond stretching* and *bending* modes. Since we can, in theory, alter every bond in a molecule, the maximum number of stretching vibrational modes for a particular type of bond is given by the number of bonds of that type in the molecule. For example, H_2O has two O–H bonds and therefore there will be two O–H stretching modes, usually denoted by the symbolism ν(O–H). In fact, the two O–H bonds do not vibrate independently, their motions are coupled so that they both move together either in or out of phase. This results in a symmetric stretching mode and an asymmetric stretching mode, which are labelled as ν_1 and ν_3 in Fig. 2.5. Similarly, there is just one bond angle defining this molecule (the 105° H–O–H angle) and so there is only one bending mode for H_2O, usually denoted as δ(H–O–H).

Question 2.3 How many Pt–Cl vibrations exist for (i) *cis*-$PtCl_2I_2$ and (ii) *trans*-$Pt(P(CH_3)_3)_2Cl_2$?

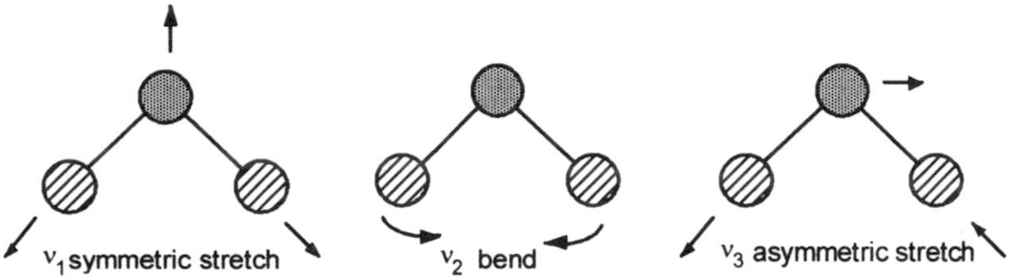

Fig. 2.5 The vibrational modes of a bent triatomic, such as water.

Vibrational modes are numbered in order of their symmetry class (most symmetric first) and from highest to lowest frequency.

For a linear triatomic such as carbon dioxide the structure is defined by two bond lengths and two bond angles and therefore there are two bond stretching modes and two bond bending modes, as shown in Fig. 2.6. The two stretching modes can be described as symmetric and asymmetric modes (ν_1 and ν_2). The two bending modes only differ by the plane in which the motion occurs and therefore the vibrational frequencies are the same. These two modes are therefore said to be *degenerate*.

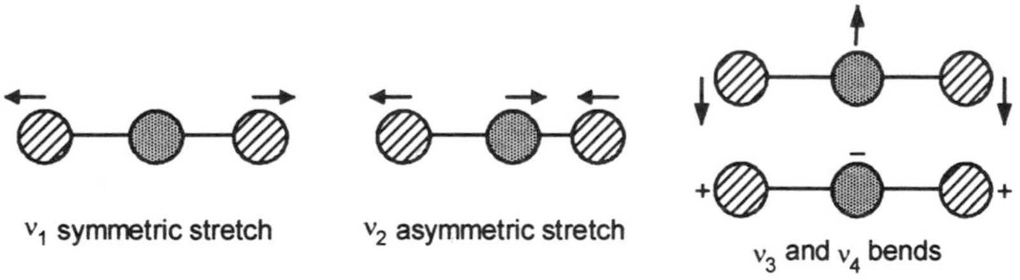

Fig. 2.6 The vibrational modes of a linear triatomic, such as carbon dioxide.

Infrared active bands

The requirement for a vibration to be infrared active (and therefore observed in an IR spectrum) is that during the vibration there must be a change in the electric dipole of the molecule. Consider the two molecules N_2 and CO and their stretching vibrations which are shown in Fig. 2.7. For dinitrogen, because the two constituents atoms are the same there is no permanent electric dipole in the molecule. As the bond stretches and contracts there will still be no electric dipole, so at no time during the vibration does the electric dipole change and so the N≡N bond stretch is not observed in infrared spectra. The same is true for all homonuclear diatomics. On the other hand, in the CO molecule there is a permanent electric dipole since oxygen is more electronegative than carbon. As the carbon–oxygen bond lengthens and shortens during the vibration the size of the dipole will change because the separation of the two charges alters. Since the electric dipole does vary during the vibration a band will be seen in the infrared spectrum arising from this stretching mode.

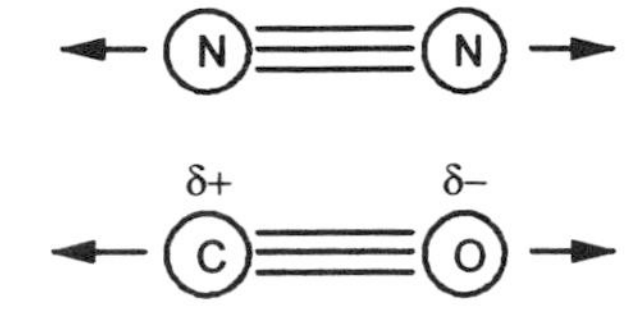

Fig. 2.7 The stretching vibrations of N_2 and CO.

In fact, the C–O stretching vibration, written as ν(CO), falls in an otherwise uncrowded part of the infrared spectrum which makes infrared spectroscopy of great use for inorganic molecules containing carbon monoxide as a ligand, such as metal carbonyls.

For example, Fig. 2.8 shows part of the infrared spectrum for tungsten hexacarbonyl, $W(CO)_6$. A single peak is observed in the carbonyl region at around 1980 cm^{-1}, which arises from the C–O stretching vibration. This is at a lower wavenumber value (and hence energy) than that observed for uncoordinated carbon monoxide (2143 cm^{-1}). In fact, the position of ν(CO) varies over a small range of frequencies from compound to compound which arises because of the way carbonyls are bound to transition metals.

Question 2.4 By considering the dipole change, how many of the vibrational modes of water (Fig. 2.5) and carbon dioxide (Fig. 2.6) are infrared active?

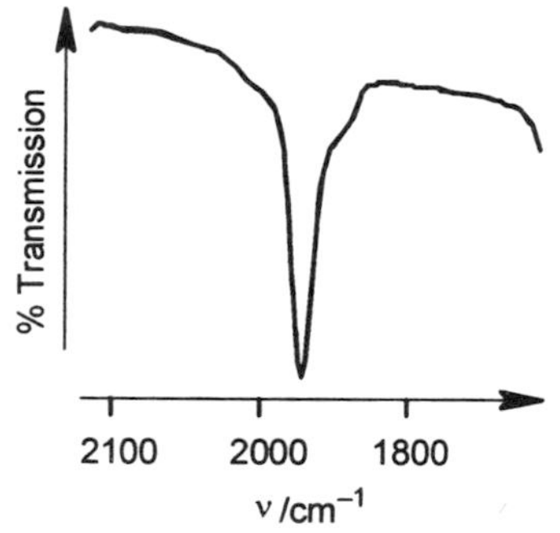

Fig. 2.8 Part of the infrared spectrum of $W(CO)_6$.

Bonding of this ligand to a metal is usually broken into two components, described by the Dewar and Chatt model. A simple molecular orbital diagram for carbon monoxide is shown in Fig. 2.9. Because oxygen has a higher atomic charge than carbon and shielding of the nuclear charge by filled orbitals is less than 100% efficient the energy levels for the oxygen 2s and 2p orbitals are lower than their carbon equivalents. This energy difference means that the $2p_z$(O) and 2s(C) atomic orbitals are the nearest in energy capable of yielding sigma-bonding and anti-bonding molecular orbitals. This results in the oxygen 2s atomic orbital remaining non-bonding as will be the carbon-based $2p_z$ orbital. This latter orbital, the highest occupied molecular orbital, or HOMO, is used to form a σ-bond to the metal.

The second part of the bonding arises from metal d-orbitals donating electron density into the LUMO (lowest unoccupied molecular orbital) of carbon monoxide, which is the π*-orbital. This part of the bonding scheme is often referred to as *backbonding* and its effect is to increase the bond order of the M–C bond, since the amount of orbital overlap increases between the metal and carbon. However, because the electron density from the metal is being donated into an anti-bonding molecular orbital of carbon monoxide, the C–O bond order is reduced. Since the energy at which vibrations occur depends on the bond strength we should expect that ν(CO) would be lower for the coordinated ligand than for free CO.

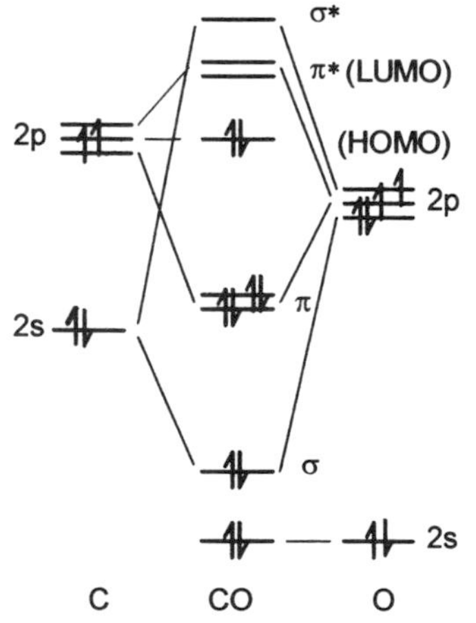

Fig. 2.9 An MO diagram for carbon monoxide.

This combination of σ-donation and π-backbonding is often referred to as a *synergic* bonding scheme, in that the two components enhance the effect of each other. Donation from the CO ligand to the metal increases the electron density on the metal which will result in an increase in the backbonding from the metal to the ligand. This in turn reduces the electron density on the metal and hence favours σ-donation from the carbonyl ligand.

(a)   (b)

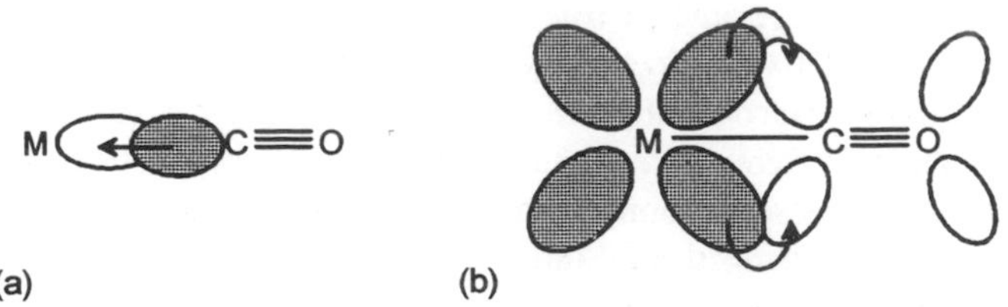

Fig. 2.10 A representation of the synergic model of transition metal carbonyl bonding.

Complex	$\nu(CO)$ (cm^{-1})
$[V(CO)_6]^-$	1860
$Cr(CO)_6$	2000
$[Mn(CO)_6]^+$	2095

Table 2.2 The IR–observed carbonyl absorption for some transition metal carbonyl species.

Question 2.5 Would you expect $Ni(CO)_4$ or $[Co(CO)_4]^-$ to possess the highest $\nu(CO)$ vibration?

We can use the position in the infrared spectrum of the C–O stretching absorption to make qualitative predictions of the amount of backbonding occurring in different complexes. For example, compare the three isoelectronic complexes $[V(CO)_6]^-$, $[Cr(CO)_6]$, and $[Mn(CO)_6]^+$; the highest formal excess of electron density, i.e. negative charge, on the metal occurs in the $[V(CO)_6]^-$ complex which means more backbonding will occur in this complex than in either of the other two. The increased backbonding will result in a decrease in the C–O bond strength and hence a decrease in vibrational frequency, as shown in Table 2.2. We should also expect the strength of the M–C bond to increase as the degree of backbonding increases; however, the change in ν(M–C) vibrations, which occur just below 400 cm^{-1}, is small, typically of the order of 10 cm^{-1}, and therefore much less easily used.

Strictly, comparisons should always be made between force constants rather than vibrational frequencies, since in that way changes in the mass of the remainder of the molecule and any structural differences are taken into account. However, comparisons between vibrational frequencies for structurally related compounds can provide useful trends.

Carbon monoxide can also act as a bridging ligand between two or more metal centres. Under these circumstances the carbonyl stretching vibration is observed at lower frequencies, typically 1850–1600 cm^{-1}. This decrease in vibrational frequency for bridging carbonyls can be rationalised if the ligand is considered to be bound as shown in Fig. 2.11. Bonding to more than one metal results in a reduced C–O bond order and lower vibrational frequency.

(a)   (b)   (c)

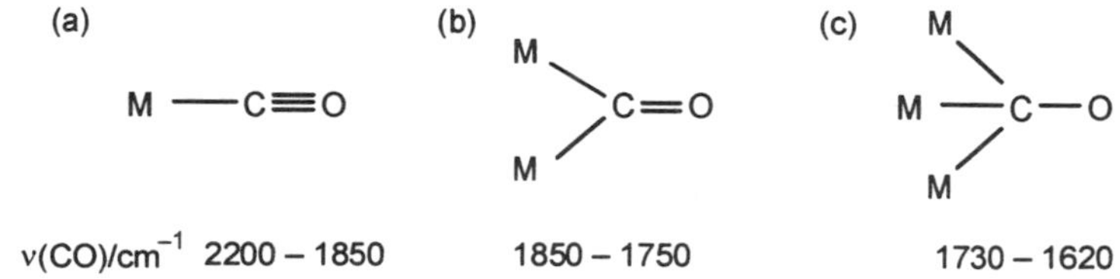

Fig. 2.11 Bonding modes of carbon monoxide and typical IR vibrational frequencies: (a) terminal; (b) bridging two metals; (c) bridging three metals.

$$M \!-\! C\!\equiv\!O \qquad\qquad \begin{matrix} M \\ \diagdown \\ \diagup \\ M \end{matrix}\!\!C\!=\!O \qquad\qquad \begin{matrix} M \\ \diagdown \\ M \!-\! C \!-\! O \\ \diagup \\ M \end{matrix}$$

$\nu(CO)/cm^{-1}$ 2200 – 1850   1850 – 1750   1730 – 1620

2.2 Symmetry

It might appear strange that the IR spectrum of $W(CO)_6$ exhibits just a single CO stretching band despite possessing six carbonyl ligands from which we might expect to see up to six carbonyl vibrations. The fact that we do not suggests that either the majority of the expected carbonyl stretching vibrations are not infrared active because there is no dipole change during the vibration, or that a number of vibrations all occur at the same energy. In reality both of these factors are partially responsible.

This molecule does not possess a permanent dipole moment, since the carbonyl ligands are evenly distributed around the central metal atom. One vibrational mode we should expect for all molecules is a symmetric stretch, which will correspond to all six of the C–O bonds vibrating in phase (written as ν_s(CO)). In this mode all C–O bonds are becoming equally long or short in phase at the same time, as shown in Fig. 2.12. The effect of this vibration is that there will be no electric dipole produced and therefore no change in dipole and so it will not be seen in the IR spectrum.

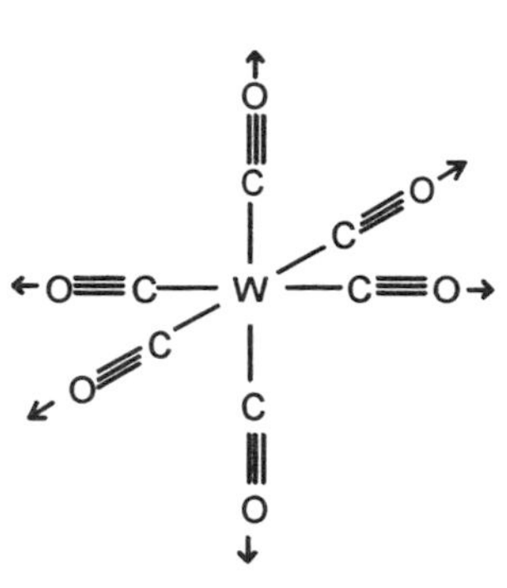

Fig. 2.12 The symmetric carbonyl stretching mode for a metal hexacarbonyl.

But what about the other modes? With some experience it is possible to make a reasonable guess of what all the vibrational modes for a molecule might be and whether they involve a dipole change. There is, however, a more reliable way and that is to make use of the symmetry of the molecule and group theory. There is not enough space here to go into this subject in detail so references to a number of suitable texts are given later.

Tungsten hexacarbonyl is octahedral and therefore belongs to the O_h point group, the character table for this group is given in Table 2.3.

These tables list, along the top, all the symmetry elements possessed by a molecule of a particular symmetry. The left-hand column lists the *Mulliken symbols* for the group, the entries in the following table give the representation for such species under each symmetry operation. The final two columns show the symmetry species for the p and d orbitals, the *x*, *y*, and *z* axes and rotations around these axes.

O_h	E	$8C_3$	$6C_2$	$6C_4$	$3C_2$	i	$6S_4$	$8S_6$	$3\sigma_h$	$6\sigma_d$		
A_{1g}	1	1	1	1	1	1	1	1	1	1		$x^2+y^2+z^2$
A_{2g}	1	1	−1	−1	1	1	−1	1	1	−1		
E_g	2	−1	0	0	2	2	0	−1	2	0		$2z^2-x^2-y^2$, x^2-y^2
T_{1g}	3	0	−1	1	−1	3	1	0	−1	−1	R_x, R_y, R_z	
T_{2g}	3	0	1	−1	−1	3	−1	0	−1	1		*xz*, *yz*, *xy*
A_{1u}	1	1	1	1	1	−1	−1	−1	−1	−1		
A_{2u}	1	1	−1	−1	1	−1	1	−1	−1	1		
E_u	2	−1	0	0	2	−2	0	1	−2	0		
T_{1u}	3	0	−1	1	−1	−3	−1	0	1	1	*x*, *y*, *z*	
T_{2u}	3	0	1	−1	−1	−3	1	0	1	−1		

Table 2.3 The octahedral symmetry point group.

To predict the symmetry of the carbonyl vibrations we consider the C–O bonds of the molecule under each of the symmetry operations and determine how many of the bonds are unchanged, that is they remain in exactly the same place in the molecule. Under the identity operation (E) all atoms, and therefore every C–O bond, remain unmoved, giving a total of six unchanged carbonyl bonds. Since all eight of the C_3 axes (which run through every pair of opposite faces of the molecule, shown in Fig. 2.13) are symmetrically equivalent we need only consider the effect of any one. Rotating the

The following molecular symmetry elements exist:
E identity operator
C_n rotation axis, by $(360/n)°$
σ_h horizontal mirror plane
σ_v vertical mirror plane
σ_d dihedral mirror plane
i centre of inversion
S_n improper rotation axis
[rotation by $(360/n)°$ followed by σ].

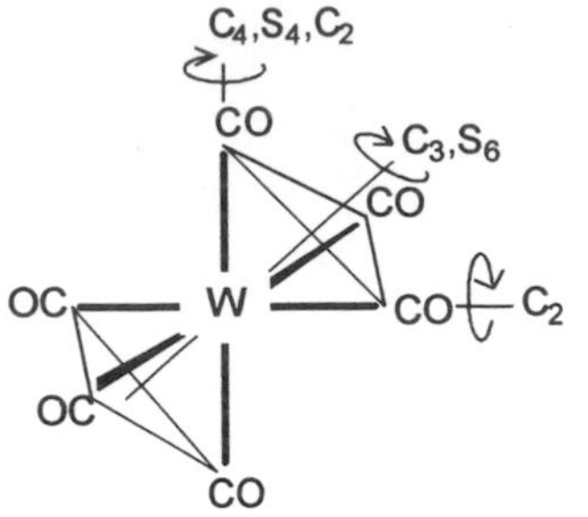

Fig. 2.13 Some of the symmetry elements of an octahedral complex such as $W(CO)_6$.

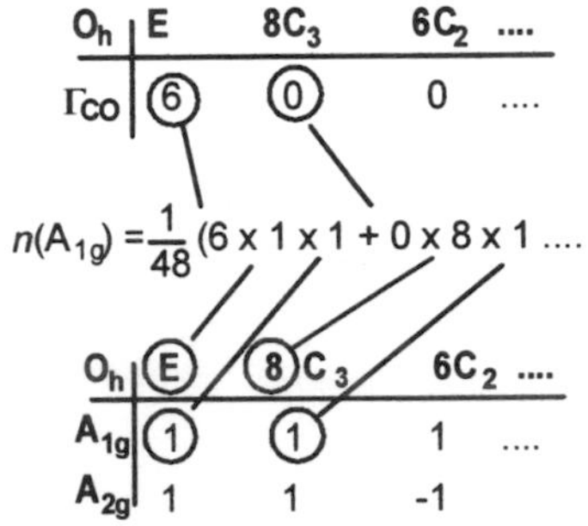

Fig. 2.14 The source of each number is as shown in the sketch for the first part of the sum.

Modes labelled A or B are singly degenerate, E doubly degenerate, and T (or sometimes F) triply degenerate.

molecule by 120° around such a C_3 axis (the angle of rotation for a C_n axis is $360/n$ degrees) results in all of the carbonyls being moved around one position. Since no C–O bonds remain unchanged, the character for this operation is zero. Going through each of the operations in turn allows us to build the entries in the table below corresponding to the *reducible representation* of the carbonyl bonds in this molecule:

O_h	E	$8C_3$	$6C_2$	$6C_4$	$3C_2$	i	$6S_4$	$8S_6$	$3\sigma_h$	$6\sigma_d$
Γ_{C-O}	6	0	0	2	2	0	0	0	4	2

However, this reducible representation corresponds to *all* the carbonyl stretching vibrational modes and it must be broken down, or reduced, into its *irreducible representation,* that is a summation of the various Mulliken symbols used in the group table (A_{1g}, A_{2g}, etc.). Only then can we determine how many carbonyl vibrations should be observed in the IR spectrum.

For relatively simple problems this process can be done by visual inspection but for larger ones it is better performed using the reduction formula given in eqn 2.4.

$$n_i = \frac{1}{h}\sum g_i \chi_i \chi_r \qquad (2.4)$$

The number of irreducible representations of type i in our reducible representation, n_i, is determined by multiplying in turn each entry for an irreducible representation (χ_i) with the number of such operations (g_i) and the corresponding entry in the reducible representation table (χ_r). The sum of these numbers is divided by the total number of operations in the group, h, which in this case is 48 [1(E)+8(C_3)+6(C_2)+6+3+1+6+8+3+6]. So to determine the number of A_{1g} irreducible representations in our reducible representation of the carbonyl bonds the calculation would be:

$$\begin{aligned} n(A_{1g}) &= \frac{1}{48}(6\times1\times1+0\times8\times1+0\times6\times1+2\times6\times1+2\times3\times1+ \\ &\quad 0\times1\times1+0\times6\times1+0\times8\times1+4\times3\times1+2\times6\times1) \\ &= \frac{1}{48}(48) \\ &= 1 \end{aligned}$$

Figure 2.14 illustrates where some of these numbers arise from. Carrying out this calculation for each of the possible representations results in an irreducible representation of $A_{1g} + E_g + T_{1u}$. Since there are a total of six carbonyl bonds we would expect a total of six vibrational modes, which is actually what we have, when we take into account that some of the modes are degenerate. The E_g representation corresponds to a doubly degenerate pair of vibrational modes and the T_{1u} to a triply degenerate set. These degenerate modes correspond to a number of different vibrations (two in the case of the E_g mode and three for T_{1u}) which occur at *exactly* the same energy and therefore appear as a single peak in the vibrational spectrum.

As stated previously, in order to observe a vibration in the infrared spectrum a change of electric dipole during the vibration is necessary. Such a dipole change occurs along the *x*, *y*, or *z* axis, or a combination of them. Therefore, irreducible representations that correspond to any of the representations of the cartesian axes will be infrared active. The symmetry label for the cartesian axes can be determined directly from the symmetry tables using the entries in the next-to-last column of the group table (see Table 2.3). The symmetry representations for the cartesian axes are denoted by (*x*, y, *z*) and the rotation of the molecule around these axes by (R_x, R_y, and R_z). For [$W(CO)_6$] the cartesian axes correspond to T_{1u} and therefore only T_{1u} vibrations will be infrared active. On the basis of this symmetry treatment we should expect to observe just a single peak in the infrared spectrum arising from vibrations of the CO bonds and this is what is observed as shown in Fig. 2.8.

These symmetry-based calculations can be carried out for a complete molecule or part of it. If one of the carbonyl ligands of a metal hexacarbonyl is replaced with an alternative donor such as bromide as in the compound $Mn(CO)_5Br$ (shown in Fig. 2.15), we can determine the number of infrared active carbonyl bands and metal–bromide stretches. In this case the molecule possesses different symmetry elements, two fourfold and one twofold rotational axis down the Br–Mn–CO axis. It has two sets of mirror planes running through the opposite equatorial bonds and between them; these are vertical and dihedral mirror planes (with respect to the C_4 axis) so it belongs to the C_{4v} point group. By considering each of the symmetry operations in turn and their effect on the C–O and Mn–Br bands separately, we obtain

Question 2.6 The bending vibrations of an octahedral molecule are T_{1u} + T_{2g} + T_{2u}. How many peaks will be seen in the IR spectrum?

Question 2.7 What will be the symmetry representation for the W–C bonds of $W(CO)_6$?

C_{4v}	E	$2C_4$	C_2	$2\sigma_v$	$2\sigma_d$
Γ_{C-O}	5	1	1	3	1
Γ_{Mn-Br}	1	1	1	1	1

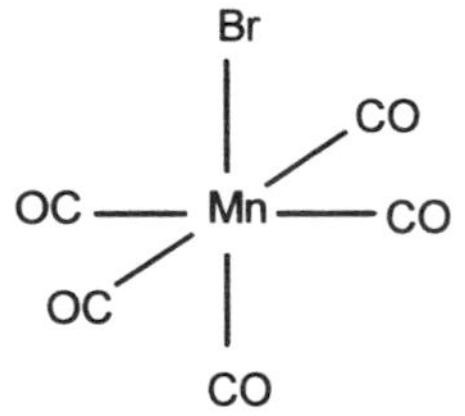

Fig. 2.15 The shape of $Mn(CO)_5Br$.

The representation for the carbonyl bonds (Γ_{C-O}) can be reduced to give an irreducible representation of $2A_1 + B_2 + E$, which corresponds to a total of five carbonyl stretching modes, two of A_1 symmetry, one of B_2 symmetry, and one doubly-degenerate E mode; this is the expected number since there are five carbonyl bonds in the molecule. Because the symmetry label for the *x* and *y* axes is E and for *z* A_1 we would expect to observe three carbonyl peaks in the IR spectrum, corresponding to the $2A_1$ and E symmetry modes. The carbonyl region of the IR spectrum for this compound is shown in Fig. 2.16 and three carbonyl bands are observed at 2140, 2051, and 2013 cm^{-1}.

By a similar process we obtain the representation for the Mn–Br bond which cannot be reduced any further since it already corresponds to the A_1 representation. This band should also be infrared active and seen in the IR spectrum. However, since the bromide ligand is much heavier than the carbonyl ligand it will vibrate at a lower frequency, at around 220 cm^{-1}, which is below the lower limit of many infrared spectrometers.

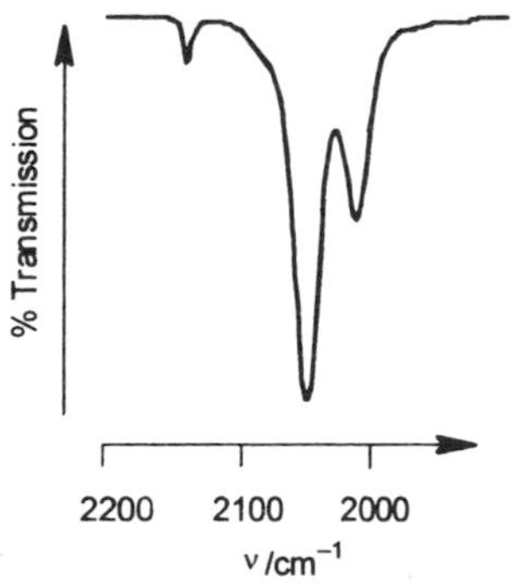

Fig. 2.16 IR spectrum of the carbonyl peaks of $Mn(CO)_5Br$ dissolved in CH_3CN.

We have only considered stretching vibrational modes so far. The symmetry of bending and torsional modes can be determined by obtaining the reducible representation for all molecular motions and removing the contributions due to translation, rotation, and bond-stretching vibrations of the molecule. However, bending modes occur at lower energies and they tend to be in a more cluttered part of the spectrum. They are therefore more difficult to assign confidently.

Table 2.4 shows the result of symmetry treatment for some of the more common molecular shapes and hence the number of infrared active stretching and bending modes to be expected.

Coordination number	Geometry	Point group	Example	Symmetry of stretching modes	Number of IR active stretches	Symmetry of bending modes	Number of IR active bending modes
2	linear	$D_{\infty h}$	CO_2	$A_{1g}^* + \underline{A}_{1u}$	1	$\underline{E}_{2u}$	1
2	bent	C_{2v}	H_2O	$\underline{A}_1^* + \underline{B}_1^*$	2	$\underline{A}_1^*$	1
3	planar	D_{3h}	BF_3	$A_1'^* + \underline{E}'^*$	1	$\underline{A}_2'' + \underline{E}'^*$	2
3	pyramidal	C_{3v}	NH_3	$\underline{A}_1^* + \underline{E}^*$	2	$\underline{A}_1^* + \underline{E}^*$	2
4	tetrahedral	T_d	CH_4	$A_1^* + \underline{T}_2^*$	1	$E^* + \underline{T}_2^*$	1
4	square planar	D_{4h}	$[PtCl_4]^{2-}$	$A_{1g}^* + B_{2g}^* + \underline{E}_u$	1	$B_{1g}^* + \underline{A}_{2u} + B_{2u} + \underline{E}_u$	2
5	trigonal bipyramidal	D_{3h}	PF_5	$2A_1'^* + \underline{A}_2'' + \underline{E}'^*$	2	$\underline{A}_2'' + 2\underline{E}'^* + E''^*$	3
5	square pyramidal	C_{4v}	IF_5	$2\underline{A}_1^* + B_1^* + \underline{E}^*$	2	$\underline{A}_1^* + B_1^* + B_2^* + 2\underline{E}^*$	3
6	octahedral	O_h	ML_6	$A_{1g}^* + E_g^* + \underline{T}_{1u}$	1	$\underline{T}_{1u} + T_{2g}^* + T_{2u}$	1

Table 2.4 The number of IR bands of some common geometric arrangements (representations underlined are IR active, Raman active modes are denoted by *).

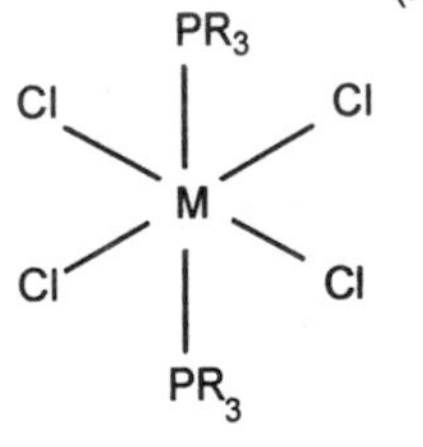

Fig. 2.17 The compound *cis*-$PtCl_2(PR_3)_2$.

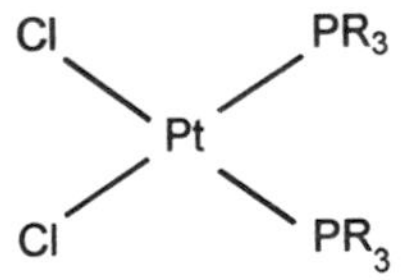

Fig. 2.18 The complex *trans*-$MCl_4(PR_3)_2$.

Question 2.8 How many IR-active M–Cl stretches should be observed for (i) *trans*-$MCl_2(PR_3)_2$ and (ii) *fac*-$MCl_3(PR_3)_3$ complexes?

The above table can be used to determine the number of vibrational bands not just for complete molecules but also for distinct parts. For example, the compounds *cis*-$PtCl_2(PR_3)_2$ adopt square-planar shapes as shown in Fig. 2.17. If we assume that the Pt–Cl and Pt–P vibrations are independent of other modes then we can consider each separately. The whole complex has C_{2v} symmetry, as do the Cl–Pt–Cl and P–Pt–P parts of the molecule. If we concentrate only on the Pt–Cl part of the molecule and effectively ignore the rest then we have a fragment similar in shape to water—that is two chloride ligands around the metal in a non-linear arrangement. We can therefore use the second entry in the above table to predict that there should be two IR-active Pt–Cl stretches. Likewise there should be two IR-active Pt–P stretching vibrations. Although this procedure tells us the number of vibrations to be expected, the Mulliken symbols given to these representations will not be correct unless the point group of the fragment coincides with that of the complete molecule.

In a similar way, rather than go through a rigorous symmetry treatment of a complete molecule such as *trans*-$MCl_4(PR_3)_2$, shown in Fig. 2.18, we can break the complex down into the equatorial square planar MCl_4 part and a separate linear $M(PR_3)_2$ section. Reference to the data in Table 2.4 suggests that we should therefore expect to observe one IR-active M–Cl stretching vibration and one IR-active M–P stretching vibration.

2.3 Infrared experiments

The basis of the IR experiment is to pass infrared radiation through a thin sample of compound and measure which energies of the applied infrared radiation are transmitted by the sample. Traditionally this was carried out using a grating or prism to scan through a range of energies produced by a source (usually an electrically heated element) and record the percentage transmission spectrum as a function of energy. It is much more likely these days that the spectrometer will be of the Fourier transform type. These work by recording the interference pattern (interferogram) that arises from two different beams of IR radiation (one fixed and one that varies) as a function of the difference between the two path lengths (in cm). Fourier transformation of this signal results in a spectrum of absorbance against energy, although it is more usual (see Chapter 1) for the energy scale to be expressed in terms of wavenumbers (cm^{-1}).

Infrared spectra can be recorded of solids, liquids, solutions, and gases using a variety of different sampling arrangements, but are probably most commonly recorded as liquids or solutions or of a solid which has been ground up with a mulling agent and pressed between two alkali halide plates. The choice of plates (see Table 2.5) and mulling agent depends on which part of the spectrum you are particularly interested in obtaining data for. One of the more common mulling agents is Nujol, a paraffin, the infrared spectrum of which is shown in Fig. 2.19.

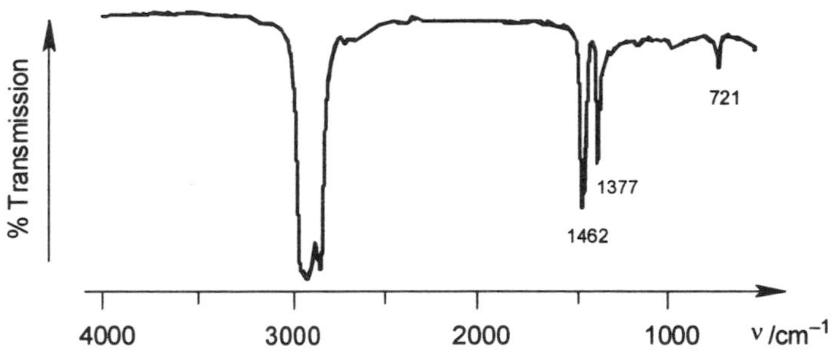

Fig. 2.19 Fourier transform infrared spectrum obtained for a sample of liquid Nujol.

Material	Spectroscopic window lower limit (cm^{-1})
NaCl	625
KBr	400
CsI	200

Table 2.5 Lower limit for some common IR plate materials.

Group frequencies

The IR spectrum shown above for Nujol contains relatively few bands with those that are present occurring in a number of well-separated regions. If spectra were to be recorded of other hydrocarbons we would find that absorptions are observed in similar regions of the spectrum. In most cases the same ligand or the same common group of atoms vibrates at very similar frequencies in a wide range of different complexes and molecules. These characteristic absorptions are known as *group frequencies* and provide one of the most straightforward methods of obtaining structural information from vibrational studies.

The concept is based on the notion that most absorptions occurring at different energies or between sets of heavy and light atoms are not coupled with other vibrations of the molecule. Simplistically, therefore, a vibration can be viewed as reflecting the atoms involved and the strength of the bond holding them together. Figure 2.20 illustrates the position of typical group frequencies for some commonly encountered groups, fragments, and linkages in inorganic and organic molecules.

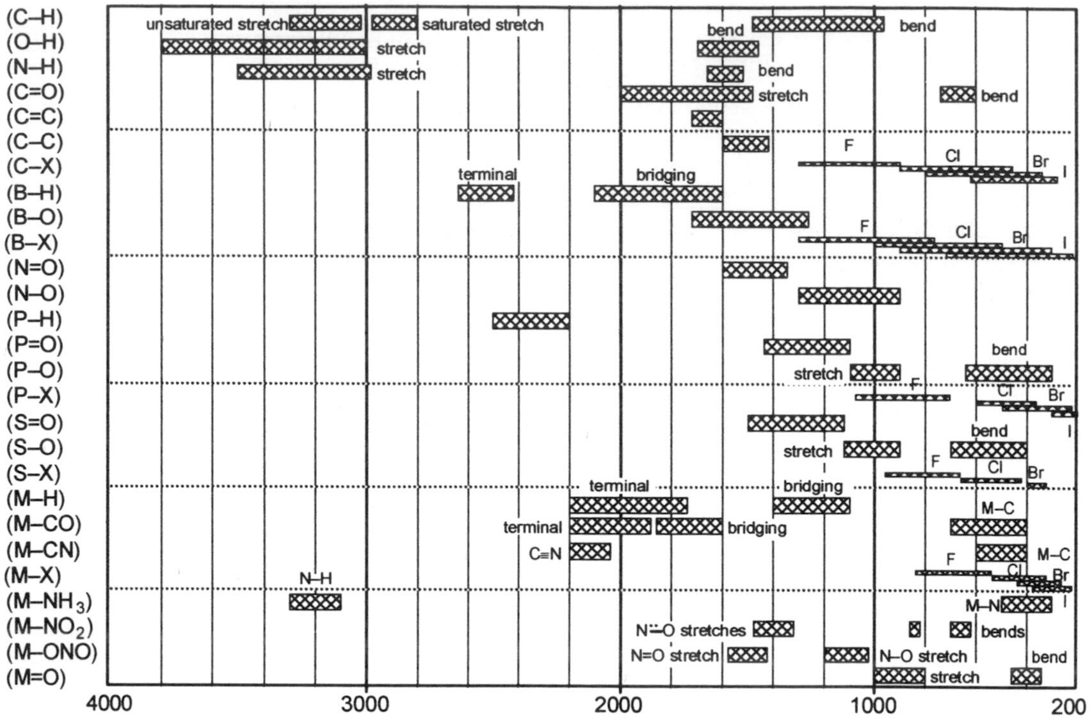

Fig. 2.20 The positions of characteristic group vibrational absorptions.

Absorption	Assignment
2925 (s)	C–H stretch
2855 (s)	C–H stretch
1462 (m)	C–C stretch
1377 (m)	C–H bend
721 (w)	C–C bend

Table 2.6 Band assignments for the IR spectrum of Nujol.

We can therefore use the data in Fig. 2.20 to assign the major peaks observed in the infrared spectrum of Nujol as follows. The peaks at 2925 and 2855 cm^{-1} correspond to the C–H stretching region and the band at 1462 cm^{-1} to the C–C stretching region. The peaks at 1377 and 721 cm^{-1} are due to bending rather than stretching vibrations. It is also common to indicate the relative intensity of each band using the letters (s) for strong, (m) for medium and (w) for weak. Additionally, (br) is used to indicate particularly broad peaks and (sh) for absorptions that appear as shoulders on another. Table 2.6 shows the complete assignment for the Nujol spectrum in this way.

It should be obvious that to record the infrared spectrum of a compound in order to determine whether it contains C–H bonds, an alternative mulling agent may need to be used since most C–H stretching absorptions fall in a similar part of the spectrum and would therefore be obscured by the C–H vibrations of Nujol. A common alternative for this purpose is hexachlorobutadiene (HCB), which displays no IR bands between 4000 and 1600 cm^{-1}.

It is possible, although it would be an extremely time-consuming process for complicated molecules, to assign all the bands in a vibrational spectrum. However, in our quest for structural information we can often just look for certain characteristic bands. For example, a number of different structures can be envisaged for the dimeric complex $[(\eta^5\text{-}C_5Me_5)_2Fe_2(CO)_4]$, two of which are shown in Fig. 2.21.

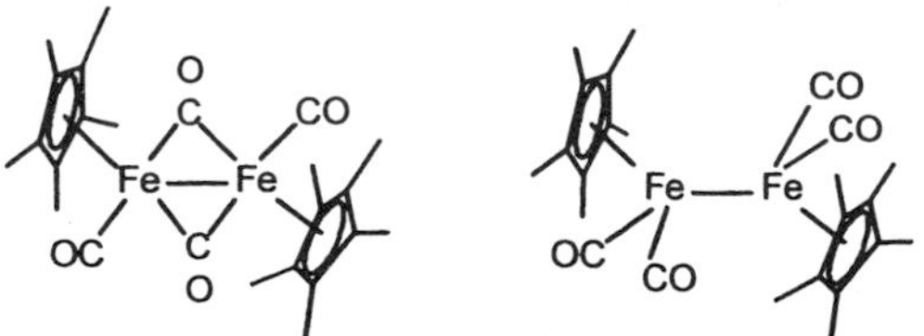

1 **2**

Fig. 2.21 Two possible structures for the complex $[(\eta^5\text{-}C_5Me_5)_2Fe_2(CO)_4]$.

The principal difference between these two structures is the presence, or otherwise, of bridging carbonyl ligands. Reference to the group frequency chart suggests that the carbonyl vibrations, but no other bands from this molecule, are likely to occur in the region 2200–1700 cm^{-1}. The Nujol mull spectrum shown in Fig. 2.22 clearly shows bands in the carbonyl region at 1960 and 1770 cm^{-1}. These can be assigned to absorptions arising from the presence of terminal CO ligands (1960 cm^{-1}) and bridging carbonyl ligands (1770 cm^{-1}). On this basis we would expect structure **1** to be adopted rather than **2**.

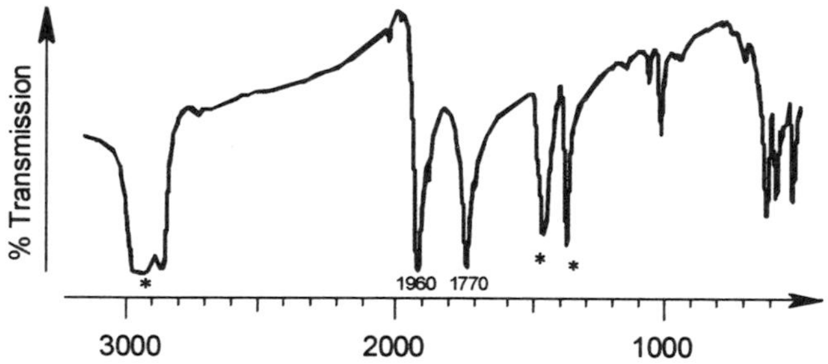

Fig. 2.22 The IR spectrum of $[(\eta^5\text{-}C_5Me_5)_2Fe_2(CO)_4]$ (bands marked * arise from the Nujol mulling agent).

As another example $SnCl_3(CH_2CH_2COOCH_3)$ reacts with 2,2'-bipyridine (abbreviated to bipy) to yield a light pink-coloured solid which according to elemental analysis corresponds to $SnCl_3(CH_2CH_2COOCH_3)$(bipy). The structure of the starting materials and the most probable products are shown in Fig. 2.23. The principal distinction between the two products **3** and **4** is whether the bipy molecule acts as a mono- or a bidentate ligand. In the latter case we would expect the ester group to no longer be coordinated to the metal. The Nujol mull IR spectrum of the product is shown in Fig. 2.24(a) and the spectrum of the starting tin-containing material is also shown, for comparison, in Fig 2.24(b).

Once confronted with a spectrum containing many different peaks, such as those shown in Fig. 2.24, it is often difficult to know exactly where to start interpreting it. We know that the sample was run in Nujol and therefore these bands can be assigned first; in the figures they are labelled with an asterisk. Next we try to focus on peaks arising from specific fragments or groups in the molecule that might be diagnostic. In the case of the starting material one of the obvious groups we might look for is the ester linkage. From the chart in Fig. 2.20 the C=O vibration should lie in the range 2000–1500 cm^{-1}. Because this group is coordinated to the metal in the starting material, by donation of electron density to the tin atom, its vibrational frequency will be at the lower end of this range, and in the spectrum the only peak observed in this range occurs at 1651 cm^{-1}.

If we look at the ν(CO) region of the IR spectrum of the product we see one peak at 1713 cm^{-1}. This C=O vibration now occurs at higher energy than that observed in the starting material which suggests that the C=O is no longer coordinated to the tin atom and that the bipy is therefore acting as a bidentate ligand as in compound **4**.

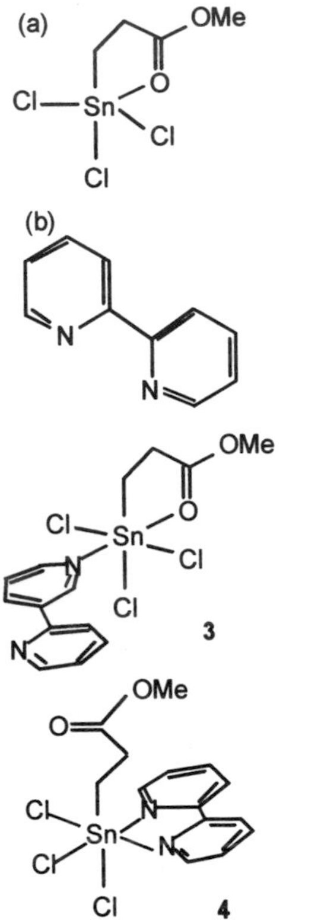

Fig. 2.23 The structures of (a) $SnCl_3(CH_2CH_2COOCH_3)$ and (b) 2,2'-bipyridine and two possible products.

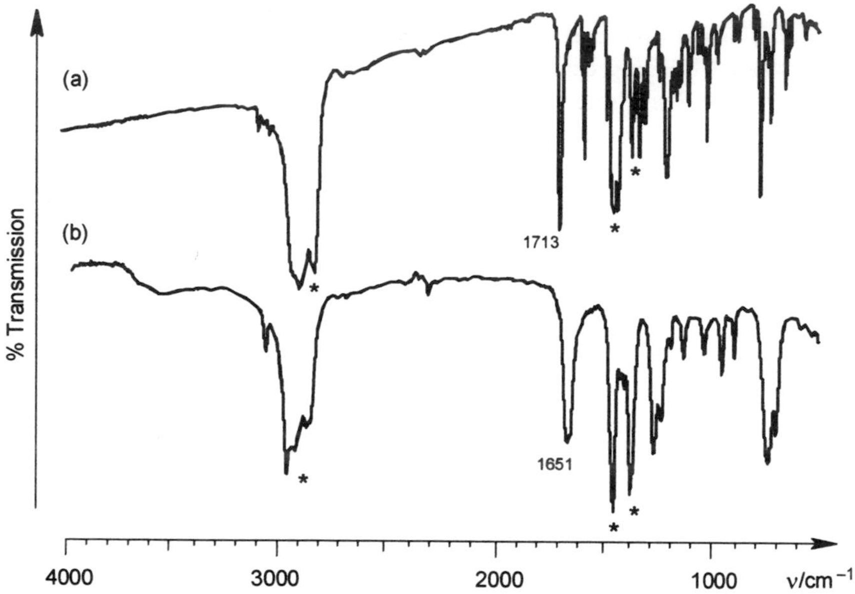

Fig. 2.24 The Nujol mull IR spectra of (a) the product after reacting 2,2'-bipyridine with $SnCl_3(CH_2CH_2COOCH_3)$ and (b) the starting complex.

Infrared spectroscopy is also very useful in determining the mode of coordination for ligands capable of coordinating through more than one atom. The compounds that arise from these differently bonded ligands are called *linkage isomers*. There are many ligands capable of this, we have already seen carbon monoxide acting as a terminal or bridging ligand. Other examples include cyanate, thiocyanate, sulfur dioxide, and the nitro group (NO_2^-). Figure 2.25 shows some of these examples and their typical vibration frequencies.

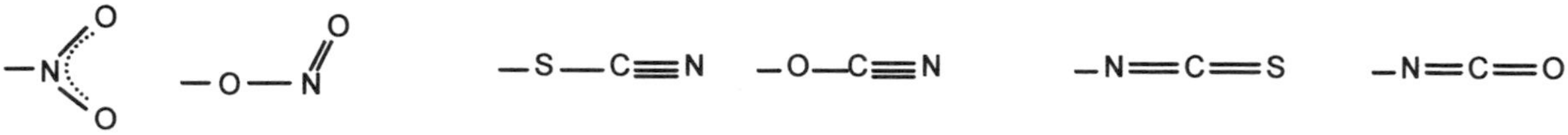

nitro-	nitrito-	thiocyanato-	cyanato-	isothiocyanato-	isocyanato-
1450–1350, 1340–1300	1480–1380, 1200–1050	2140–2100, 720–680	2210–2000, 1300–1150	2100–2040, 850–800	2250–2150, 1450–1300

Fig. 2.25 Some of the common ligands capable of forming linkage isomers.

The NO_2^- group can coordinate to a metal centre in a variety of ways including through the nitrogen atom, when it is called the nitro ligand, or through one of the oxygen atoms, referred to as the nitrito ligand. In the first case the charge is delocalised resulting in two N–O bonds formally of bond order 1.5. We would expect two N–O absorptions because there are two N–O bonds. Because the bonds are similar they will couple to give a symmetric and an asymmetric stretching vibration with similar frequencies. In the case of the nitrito ligand one N–O bond can be viewed as a single bond and the other as a double bond which means we would expect to see one higher vibrational frequency and one lower one.

So far the spectra presented have been of mulled solids or solutions, but it is also quite common for spectra to be recorded of gases. However, gas-phase spectra frequently show additional structure on the observed bands arising from rotational excitation as well as vibrational transitions. These additional features have not been observed so far because rotation is hindered in the solid and solution phases. Figure 2.26 shows an idealised representation of the fine structure observed on some infrared bands. Three distinct sets of peaks are observed: a single central peak corresponding to vibrational transitions which are not accompanied by a change in rotational energy level and on either side a set of bands in which there is a change in rotational quantum number by either +1 or −1. The actual form of the fine structure, in particular whether the central band labelled as the Q band is observed or not, depends on the form of the vibration and the shape of the molecule. For the majority of small molecules vibrations in which the dipole change occurs parallel with the principal rotational axis (usually the highest order symmetry axis) will show P and R but not Q fine-structure, because in addition to the vibrational selection rule of $\Delta v = \pm 1$ there is a rotational selection rule based on the rotational quantum number (J) of $\Delta J = \pm 1$. For those vibrations where the dipole change occurs perpendicular to the rotational axis, the rotational selection rule becomes $\Delta J = 0, \pm 1$ and so these absorptions should show P, Q, and R fine-structure.

Question 2.9 The compound $K_4[Ni(NO_2)_6]$ exhibits bands at 1387, 1347, 1325, and 1206 cm^{-1}. What does this suggest about the bonding of the NO_2 ligands?

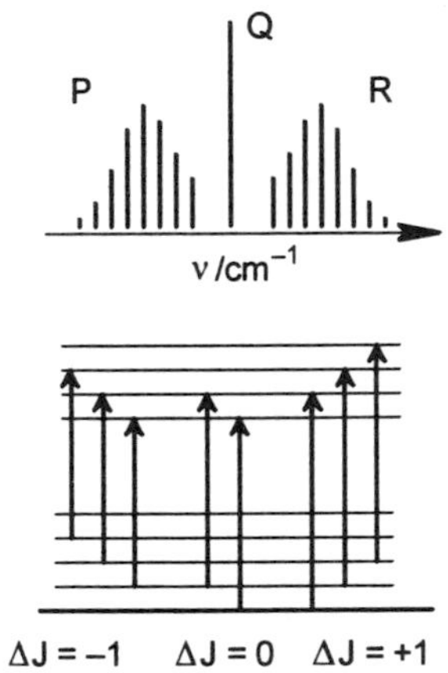

Fig. 2.26 An idealised representation of the PQR structure observed for some IR bands and their source.

The infrared spectrum of a sample of a gas mixture containing sulfur dioxide and carbon dioxide is shown in Fig. 2.27. Although the resolution of the spectrum is not sufficiently high to see the individual peaks arising from the rotational fine structure, all five of the observed peaks, labelled A to E, appears as more than a single absorption due to the presence of rotational structure. The first peak, labelled A, shows P and R structure similar to that in Fig. 2.26 whilst the absorption marked B, which is shown in greater detail in Fig. 2.28, exhibits PQR structure. The two peaks C and E both appear to show PR structure and the peak at around 662 cm^{-1} marked D is dominated by a very intense sharp Q-band absorption with much less intense P and R bands.

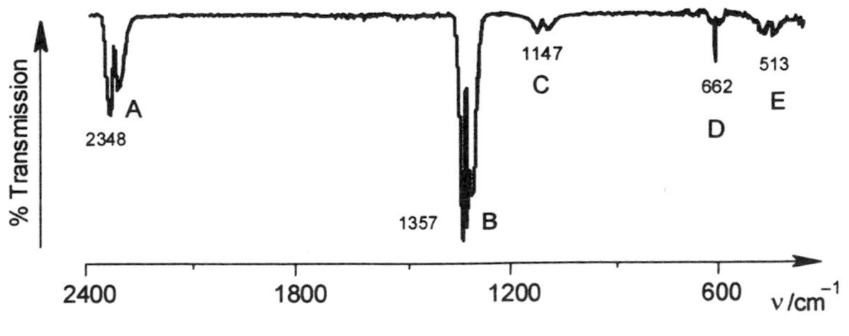

Fig. 2.27 The gas-phase IR spectrum of a sample of CO_2 and SO_2.

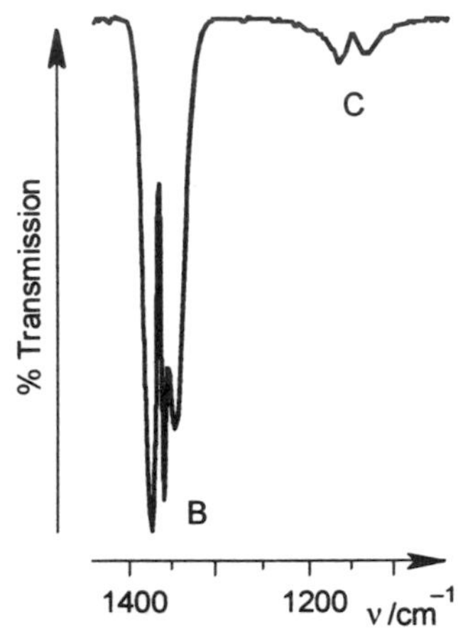

Fig. 2.28 Expansion of Fig. 2.27 in the range 1450–1050 cm^{-1}.

Carbon dioxide is a linear molecule and so will possess a total of four (3N–5) vibrations whereas sulphur dioxide is a bent triatomic molecule and will therefore exhibit three vibrational modes. Application of symmetry, or reference to Table 2.4, shows that for SO_2 there are three separate modes two of which correspond to stretches (A_1 + B_1) and one is a bending mode (A_1). For CO_2 two stretching modes are expected, of which one, the A_{1g}, is infrared inactive, and a single infrared active doubly degenerate bending mode of E_{2u} symmetry. We should therefore expect a mixture of these two gases to give rise to a total of five infrared absorptions.

The next stage is to try and assign the bands on the basis of group frequencies. From the data in Fig. 2.20 we should expect the S=O stretches to be within the range 1500–1100 cm^{-1} and the C=O stretches around 2000 cm^{-1}. On this basis we can assign the bands at 1357 and 1147 cm^{-1}, marked B and C, to the ν(S=O) and absorption A at about 2348 cm^{-1} must be ν(CO). According to the group frequency chart the bending mode for carbon dioxide is in the region 750–600 cm^{-1} and the O=S=O bending vibration at slightly lower energy (as might be expected since sulfur is heavier than carbon) in the range 700–400 cm^{-1}. The band at 662 cm^{-1} marked as D is, therefore, most likely the bending mode for CO_2 and band E (553 cm^{-1}) the bending mode for SO_2. Thus, on the basis of the expected number of absorptions and typical group frequencies we have assigned all the peaks in the spectrum to the appropriate species.

There is one final assignment to try to make and that is to identify which of the peaks at 1357 and 1147 cm^{-1} corresponds to which of the two S=O stretching modes of sulfur dioxide. Since SO_2 is a bent triatomic its three vibrational modes will have the same form as those for other similarly bent triatomic molecules such as water, which are shown in Fig. 2.5. During the symmetric stretch, labelled as ν_1, the change in dipole is parallel with the principal axis so no Q band should be seen in the rotational fine structure. But during the asymmetric vibration the dipole change is not parallel to the rotational axis and so a Q-band should be observed in conjunction with this vibrational absorption.

Peak	Rotational structure	Assignment
2348(m)	PR	$\nu_{as}(CO_2)$
1357(s)	PQR	$\nu_{as}(SO_2)$
1147(w)	PR	$\nu_s(SO_2)$
662(w)	PQR	$\delta(CO_2)$
573(w)	PR	$\delta(SO_2)$

Table 2.7 Assignment of the peaks observed in Fig. 2.27.

Figure 2.28 shows an expansion of these two bands from which it is obvious that although the individual peaks of the fine structure are not visible, the band at higher frequency clearly possesses a central Q-band feature whilst the lower frequency band does not. The band at 1357 cm^{-1} must therefore correspond to the asymmetric S=O stretching mode and the much weaker peak (because there is a much smaller overall change in the dipole during the vibration) at 1147 cm^{-1} corresponds to ν_s(S=O). A complete assignment of this spectrum is given in Table 2.7.

Fingerprints

A further use of vibrational spectroscopy is to confirm the identity of a compound by comparison with a known sample. In this instance rather than looking for a limited number of peaks specific to a particular part of the molecule, IR spectra are recorded for both the unknown and known compounds and the two spectra are then compared either by eye or using a computer. For the compounds to be confirmed as the same, all peaks should match in terms of position and relative intensity. The peaks at higher energy (4000–1000 cm^{-1}) are typical of certain characteristic groups and ligands within molecules, while those at lower energies (typically below 1000 cm^{-1}) are known as the *fingerprint* region since they are sensitive to small variations between similar compounds. Because of this distinction it is not usual to try to assign fingerprint peaks unless you have a very good idea of the structure of the compound.

Analysis of mixtures

An extension of the fingerprint application is to use infrared spectroscopy as a method of identifying a mixture of compounds. Since the spectrum recorded for a mixture will be a weighted composite of the individual spectra, this is an area where computer subtraction is of great use. By comparison of the spectrum with a library of standard spectra the identification and composition of complex mixtures can be determined.

There are a number of compendia of infrared spectra, both printed and in computer libraries, available for comparison, references to some of which are given at the end of this chapter.

2.4 Raman spectroscopy

The second common form of vibrational spectroscopy is based on a different physical process. When light of energy less than that required to promote a molecule into an excited electronic state is absorbed by a molecule a *virtual excited state* is created. This virtual state is of very short lifetime and the majority of the light is re-emitted over 360° at the same energy: this is called *Rayleigh scattering*. However, C.V. Raman noticed, in 1928, that the energy of a small proportion of the re-emitted light differs from the incident radiation (ν_0) by energy gaps that correspond to some of the vibrational modes. The non-Rayleigh light occurs both at higher energy ($\nu_0 + h\nu$), where the energy difference derives from a loss of vibrational energy of the molecule, and at a lower energy ($\nu_0 - h\nu$), because a vibrational promotion has occurred. These two events are shown schematically in Fig 2.29.

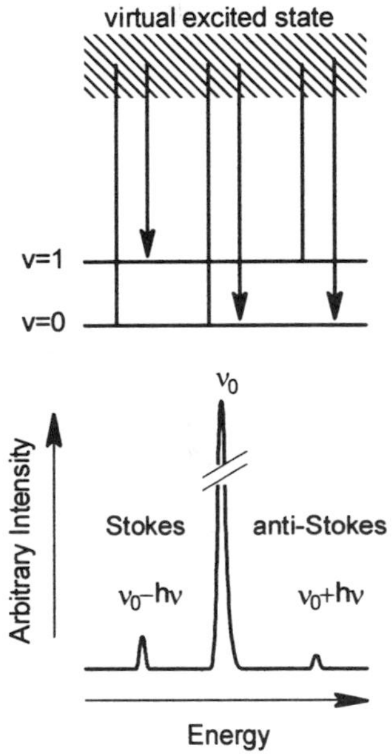

Fig 2.29 Energy levels involved in Raman spectroscopy

The ground vibrational state is more densely populated than the excited state, and it is therefore more likely that promotion of a ground state molecule into a vibrationally excited state will occur rather than the reverse. The peaks observed at lower frequency than the incident radiation (called Stokes lines) will, by virtue of the greater population and hence greater probability, be more intense than those derived from excited vibrational states at higher energy, called anti-Stokes lines.

Unfortunately, the Raman effect is very weak and a lot of monochromatic radiation is required so that there is sufficient of the energy-shifted light available to be detected. This need and the requirement for a well-collimated and monochromatic source of light mean that the radiation in Raman experiments is usually derived from a laser. A schematic representation of the instrument set-up is shown in Fig. 2.30.

However, problems may arise for dark-coloured samples which absorb an appreciable amount of the incident light resulting in localised heating and ultimately burning of the sample. There may also be problems with samples that are fluorescent, since this will often swamp the Raman effect. These problems have been reduced by the advent of Fourier-transform Raman (FTR) spectrometers which use an infrared laser to overcome fluorescence problems and interferometry techniques to increase the speed of acquiring spectra.

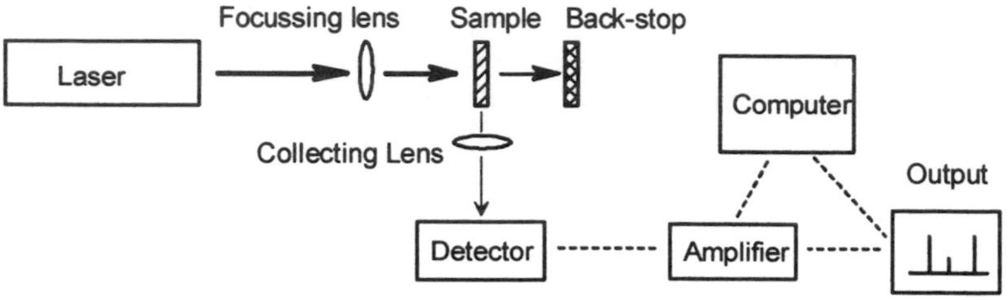

Fig. 2.30 A schematic representation of a Raman spectrometer.

The information obtained from Raman spectroscopy is vibrational frequencies, measured as a Raman shift relative to the exciting energy source. But because the physical processes involved are different from those in IR spectroscopy the selection rules also differ. Instead of requiring a change in electric dipole for the vibrational mode to be observed, as in IR spectroscopy, a change in the *polarisability* of the molecule during a vibration is required for Raman activity. The polarisability is a measure of the ease with which the electron cloud may be distorted, or polarised.

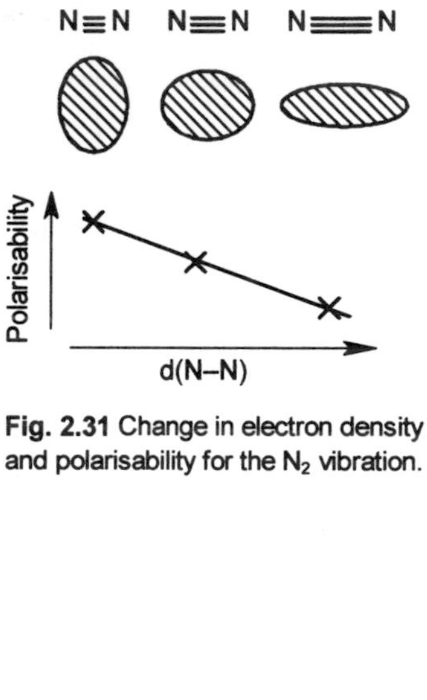

Fig. 2.31 Change in electron density and polarisability for the N_2 vibration.

During the course of the N≡N stretching vibration of dinitrogen the electron distribution will change, as illustrated in Fig. 2.31. When the N–N bond is at its most stretched the electron density will be spread more thinly over a larger volume and can be distorted more easily than if it is concentrated in a very small area. Because the electron density is altered in such a way that further distortion, or polarisation, varies between the extremes of the vibration, the band will be Raman active.

In general, it is not easy to determine whether the polarisability changes for the vibrational modes of more complicated molecules. But, just as it is possible to determine the activity of vibrational bands for a particular molecule in the infrared using group theory, so the Raman activity may also be obtained this way. In this instance, if the representation of a particular vibrational mode corresponds to one or more of the 'squared' terms, e.g. x^2, y^2, z^2 or xy, xz, yz or combinations of these, then it will be Raman active.

The infrared activity of the carbonyl stretching vibrations of $W(CO)_6$ were determined earlier. We can now determine whether any of them will be Raman active. The symmetries of the carbonyl stretching modes were determined as $A_{1g} + E_u + T_{1u}$. The final column of the group table (Table 2.3) gives 'squared' terms corresponding to the symmetry labels A_{1g}, E_g, and T_{2g}. For this compound, therefore, only the A_{1g} carbonyl stretch will be Raman active.

Raman applications

One area of study where the Raman effect is frequently more useful than infrared spectroscopy is for the determination of low frequency vibrations (below *ca.* 400 cm^{-1}), for example the stretching frequency of compounds containing heavy elements or of weakly bound molecules. This is because IR studies are usually carried out as thin samples between alkali-metal halide plates which start to absorb strongly at low energies, see Table 2.5, and so mask any sample absorptions in these regions. However, because the Raman spectrum may be recorded using monochromatic light, any material which is clear to visible light can be used as a sample holder, for example thin-walled glass tubes. Figure. 2.32 shows part of the Raman spectrum of the product of the reaction of PMe_3 with I_2 in diethyl ether and the structure of this compound.

The Raman spectrum clearly shows two peaks, the lower energy of which (at *ca.* 210 cm^{-1}) is assigned as the I–I stretching mode. By comparison, ν(I–I) for diiodine occurs at 215 cm^{-1} which suggests that the I–I bond in the adduct is weaker than in diiodine. This is consistent with the donation of electrons from the lone pair of the phosphine into an orbital of diiodine possessing anti-bonding character resulting in a decrease in the I–I bond order and a lower vibrational frequency.

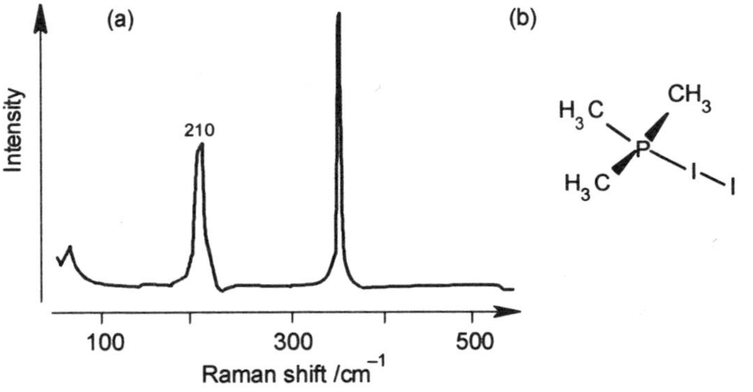

Fig. 2.32 (a) Raman spectrum of the PMe_3–I_2 adduct and (b) its solution-phase structure.

Depolarisation measurements

Depending on the set-up of a Raman spectrometer it may be possible to measure the scattered light in two different planes and hence obtain a value for the *depolarisation ratio*. If light is directed onto a sample along, say, the *x* direction and the detected light is along the *y* direction then the light will be polarised in the *z* direction such that the depolarisation ratio is given by

$$\rho = \frac{I_z(\perp)}{I_z(\parallel)}$$

For all non-totally symmetric vibrations depolarised lines are observed with $\rho = 3/4$. In the case of totally symmetric modes bands will possess a depolarisation ratio in the range $0 < \rho < 3/4$. By placing a polariser before the detector and measuring the peak intensities in two perpendicular planes it is possible to identify which bands correspond to totally symmetric modes.

Question 2.10 The following Raman shifts (and depolarisation values) are obtained for a sample of CCl_4: 218 cm^{-1} (0.75), 314 cm^{-1} (0.75), 459 cm^{-1} (0.02). Assign the peaks as far as possible.

Complementary nature of Raman and IR spectroscopy

We have seen that we would expect to observe ν(N≡N), the N–N stretch of dinitrogen, in the Raman spectrum because it results in a change in the polarisability of the molecule. Conversely, this vibration is infrared inactive because there is no associated change in electric dipole. Similarly, one of the carbonyl stretching vibrations of $W(CO)_6$ is infrared active and a different one is Raman active. Indeed, for all molecules such as these, possessing a centre of symmetry (symmetry operation i), vibrational modes which result in a change in electric dipole and hence are infrared active do not give rise to a change in polarisability and so are Raman inactive and vice versa. In this respect Raman and infrared spectroscopies are complementary and can be used as a method to infer the presence, or otherwise, of a centre of symmetry in a molecule. Compounds which possess a centre of symmetry will not display any common peaks in their Raman and IR spectra. This is termed the *mutual exclusion principle*.

Question 2.11 The IR spectrum of KrF_2 has bands at 233 and 588 cm^{-1}, the Raman spectrum displays a single band at 449 cm^{-1}. Is KrF_2 linear or bent?

There are other characteristics that make IR and Raman spectroscopic techniques complimentary. Generally, polar bonds which result in a large dipole absorb strongly in the infrared, while covalent bonds which are more easily polarised absorb more strongly in Raman spectra. This has important implications when choosing solvents for vibrational spectroscopic work. For example, water is a poor Raman scatterer and therefore aqueous solutions are more amenable to study by Raman than IR spectroscopy.

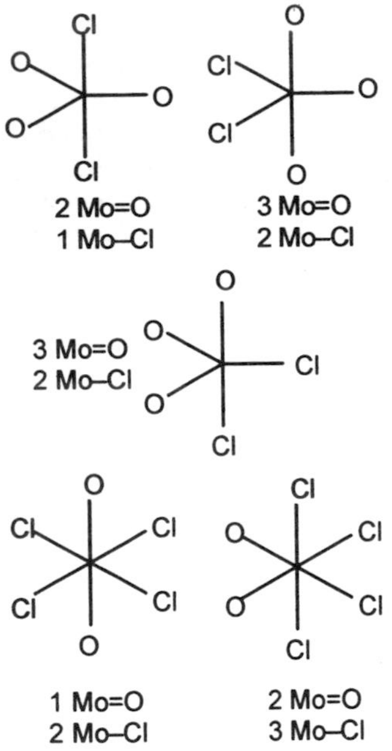

Fig. 2.33 Predicted Mo=O and Mo–Cl Raman activity of the complexes $[MoO_3Cl_2]^{2-}$, $[MoO_2Cl_4]^{2-}$.

The Raman spectrum of sodium molybdate (Na_2MoO_4) dissolved in HCl exhibits peaks at 964, 925, 392, 311, 246, and 219 cm^{-1} with the bands at 925 and 311 cm^{-1} being polarised. What can be deduced from these data?

From a chemical standpoint we should expect these conditions to result either in protonation or chlorination of the molybdate species. Firstly, the negative evidence—there are no bands corresponding to ν(Mo–H) or ν(O–H) observed in the spectrum, so protonation does not appear to have occurred. The peaks at 964 and 925 cm^{-1} can be assigned to ν(Mo=O) vibrations, and because there are two of these bands there must be at least two Mo=O bonds in the molecule. It is more difficult to assign the remaining bands unambiguously since both Mo–Cl stretching vibrations and Mo=O bending modes are expected to fall within a similar region of the spectrum. In other oxides of molybdenum the Mo=O bending modes are found between 370 and 420 cm^{-1} and on this basis the band at 392 cm^{-1} is assigned to such a mode. The remaining three bands are tentatively assigned as Mo–Cl stretching modes. The most likely products would arise from substitution of oxide for chloride ligands leading to the complexes $[MoO_3Cl_2]^{2-}$, $[MoO_2Cl_4]^{2-}$, and $[MoOCl_6]^{2-}$. The last of these can be discounted since it would only give rise to one ν(Mo=O) peak.

A symmetry treatment of the possible isomers for the other complexes can be used to determine the number of Raman-active stretching vibrations resulting in the data shown in Fig. 2.33. Only for one structure is there agreement between the observed and expected number of Mo=O and Mo–Cl stretching modes. On this basis it would appear that *cis*-$[MoO_2Cl_4]^{2-}$ is the species present in solution.

2.5 Potential problems

The following table identifies some of the more common problems encountered in IR spectroscopic studies and suggests possible remedies.

Problem	Possible cause	Suggested remedy
More peaks than expected	(a) Mixture of compounds (b) Lower molecular symmetry than expected (c) Partially resolved rotational structure	(a) Effect separation if possible (b) Re-evaluate spectrum
Fewer peaks than expected	(a) Molecule possesses higher symmetry than expected (b) Accidental overlap of peaks (c) Very low IR activity for some bands	(a) Re-evaluate spectrum (b, c) Record Raman spectrum
Very broad peaks	(a) Poorly mulled sample (b) Hydrogen bonding	(a) Grind sample well before adding mulling agent (b) Try solution-phase spectra

2.6 Further questions

2.12 The IR spectra of $(\eta^6\text{-}C_6H_6)Cr(CO)_3$ and $(\eta^6\text{-}C_6H_3Me_3)Cr(CO)_3$ dissolved in hexane exhibit bands at 1984, 1917 cm^{-1} and 1971, 1903 cm^{-1} respectively. What conclusions can be drawn about the bonding of the two arene ligands?

2.13 Which vibrational spectroscopic method would be most appropriate in order to study the following systems: (i) K_2PdBr_4; (ii) O_2; (iii) very dilute samples? Explain why.

2.14 The IR spectrum of $(Me_3PO)_2SnF_4$ exhibits bands due to stretching modes at 1083, 1065, 577, 561, 551, 472, and 448 cm^{-1}. Assign these bands as far as possible and determine the geometry of the complex.

Further reading

G. Davidson, Group theory for chemists, MacMillan Education Ltd (1991).
F.A. Cotton, Chemical applications of group theory, Wiley (1990).
K. Nakamoto, Infrared and Raman spectra of inorganic and coordination compounds, Wiley (1986).
E.B. Wilson, J.C. Decius and P.C. Cross, Molecular vibrations, McGraw-Hill, New York (1945).
The Aldrich library of FT-IR spectra, Aldrich Chemical Company Inc., Wisconsin, (1985).

3 Resonance spectroscopy

3.1 Nuclear magnetic resonance (NMR)

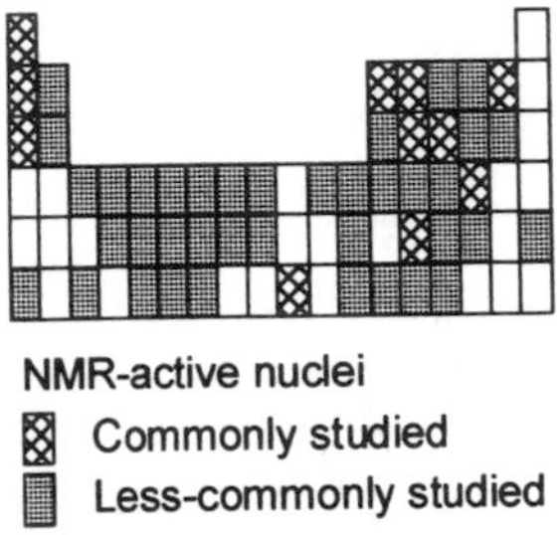

Fig. 3.1 NMR active nuclei in the periodic table.

There can be little doubt that nuclear magnetic resonance (NMR) spectroscopy is probably the single most widely used and important physical technique available to the modern practical chemist. This is because of its wide applicability, its relative ease of use, and the amount of chemical and structural information that can be obtained from its use. The technique is frequently associated only with proton and carbon nuclei but there are many other elements to which it can potentially be applied, as shown in Fig. 3.1. However, for a number of reasons, some of which will be mentioned later, not all of these nuclei are equally suitable for study.

Basic principles

A nucleus will possess a magnetic moment when its spin quantum number, I, is non-zero. In the absence of a magnetic field all the magnetic states of that nucleus will be degenerate. However, this is not the case if the nucleus is placed in an external magnetic field when a number of allowed orientations, given by eqn 3.1, will be adopted.

$$m = I, I-1, I-2, \ldots, 0, \ldots, -I+1, -I \qquad (3.1)$$

This results in a total of $(2I + 1)$ possible orientations, from which transitions resulting in a change of $\Delta m = \pm 1$ are allowed. It is this splitting of the otherwise degenerate energy levels in a magnetic field that makes NMR spectroscopy possible.

The most widely studied nuclei (^1H, ^{13}C, ^{19}F, ^{31}P) all possess a spin quantum number of ½, and so when these nuclei are placed in a magnetic field two spin states arise ($m = +½$, $m = -½$). A single transition between these two energy levels is possible, as shown in Fig. 3.2. The difference in energy between these two states is given by

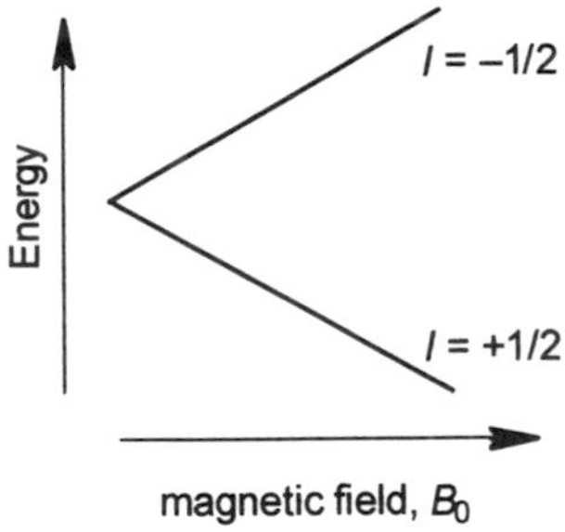

Fig. 3.2 Energy levels for a spin $I = ½$ nucleus in an externally applied magnetic field.

$$\Delta E = \frac{\gamma h B_0}{2\pi} \qquad (3.2)$$

where B_0 is the strength of the applied magnetic field, and γ the gyromagnetic ratio for the nucleus under study (see Table 3.1). The frequency of radiation that corresponds to this energy is called the *resonance frequency* and is given by eqn 3.3

$$2\pi\nu = \gamma B_0 \qquad (3.3)$$

For the type of magnetic field strengths usually applied (2–14 tesla) this corresponds to radiation within the radio-frequency part of the electromagnetic spectrum.

The ratio of the equilibrium populations in the upper and lower states is related to their energy difference and is given by the Boltzmann distribution. As we saw earlier, for a set of protons which resonate at 200 MHz at room temperature, the population ratio n_{upper}/n_{lower} is 0.999975, that is a difference of about 1 in 10,000. Because of the small difference in populations the NMR technique, from a spectroscopic point of view, is a relatively weak effect.

Chemical shift

Even though a magnetic field of B_0 is applied to a sample, the effective magnetic field (B_{eff}) experienced by the nucleus will not necessarily be the same, since the motion of the electrons surrounding the nucleus will result in induced magnetic fields. The nucleus is therefore said to be *shielded* from the external field by its surrounding electrons,

$$B_{eff} = B_0(1-\sigma) \qquad (3.4)$$

where σ is the *shielding constant.* The shielding constant will vary as the local environment changes. This will result in the energy and hence frequency, that need to be applied to cause a transition, altering, and therefore the resonance frequency is characteristic of the environment around the nucleus.

The effects of the shielding constant are small and since the absolute value of σ is not usually required it is more usual for the difference between the resonance frequency of the nucleus in the unknown (ν_{obs}) and in a standard, or reference, compound (ν_{ref}) to be determined. This difference is very small so is multiplied by a million. The frequency will also vary depending on the strength of the magnetic field so this quantity is divided by the frequency for the *standard compound* to provide the δ or ppm (parts per million) scale which is a dimensionless quantity independent of the applied magnetic field.

$$\delta = 10^6 \times \left(\frac{\nu_{obs} - \nu_{ref}}{\nu_{ref}}\right) \qquad (3.5)$$

Since the chemical shift of the reference compound is defined as 0 ppm we can convert between the chemical shift (ppm) and frequency (Hz) scales and vice versa using the relationship

$$\delta = 10^6 \times \frac{\Delta\nu}{\text{spectrometer frequency}} \qquad (3.6)$$

For example, two peaks separated by 0.5 ppm on a spectrometer operating at 200 MHz correspond to a resonance frequency difference of:

$$\Delta\nu = \frac{(\delta \times \text{spectrometer frequency})}{10^6}$$
$$= \frac{0.5\ \text{ppm} \times 200\ \text{MHz}}{10^6}$$
$$= 100\ \text{Hz}$$

Question 3.1 Two peaks in a proton NMR spectrum recorded at 100 MHz occur at 4.1 and 4.2 ppm. What is their separation in Hertz?

By convention, every nucleus has an independent reference compound used to define 0 ppm. For most nuclei this is a single internationally accepted reference compound. These are usually chosen because they are chemically stable and inert and possess one, or preferably more, nuclei in well-defined chemical environments. Some of the more commonly studied spin-½ NMR nuclei are listed in Table 3.1 along with their reference compounds and other relevant data. Corresponding details for nuclei with a spin greater than ½ are given in Table 3.5.

Different nuclei have different magnetogyric values (γ) so different frequencies are needed for the nuclei to resonate, as defined by eqn 3.3. These resonant frequencies are usually separated by many MHz whilst the differences arising from chemical shifts are in the order of Hertz so there is little chance that the NMR spectrum of one nucleus will overlap or interfere with that of a second nucleus.

Nucleus	Natural abundance (%)	Relative NMR frequency (MHz) (B_0 = 4.7 T)	Magnetogyric ratio (10^7 T^{-1} s^{-1})	Receptivity relative to ^{1}H	Standard reference compound	Common range (ppm)
^{1}H	99.99	200.0	26.75	1.0	$(CH_3)_4Si$	−30 – 20
^{13}C	1.11	50.2	6.73	0.016	$(CH_3)_4Si$	−100 – 400
^{19}F	100.0	188.2	25.18	0.83	$CFCl_3$	−200 – 200
^{29}Si	4.7	39.8	−5.32	0.0078	$(CH_3)_4Si$	−350 – 40
^{31}P	100.0	81.0	10.84	0.066	85% aq. H_3PO_4	−100 – 250
^{77}Se	7.58	38.2	5.101	0.0005	Me_2Se	−300 – 200
^{119}Sn	8.58	74.5	−10.03	0.052	Me_4Sn	−1000 – 8000
^{195}Pt	33.8	43.0	5.83	0.0099	$[Pt(CN)_6]^{2-}$	−200 – 15000

Table 3.1 NMR properties of some commonly encountered spin ½ nuclei.

By considering eqns 3.3 and 3.4 it should be clear that when the degree of shielding, σ, increases then B_{eff} decreases and hence the resonance frequency, ν, decreases. This in turn means that a peak is shifted to a lower chemical shift, or ppm, value. Conversely, if a peak shifts to a more positive ppm value the degree of shielding of the nucleus under study has decreased, in other words the nucleus is more *deshielded*. For example, the position of the proton NMR signals for HI, HBr, HCl, and HF increase from −13.5 to 1.9 ppm which means that the proton in HF is more deshielded compared with the proton in HI. This is as expected since fluorine is more electronegative than iodine and will therefore result in a greater decrease in electron density around the proton in HF when compared with the proton in HI.

It is possible to describe the relative position of two peaks in an NMR spectrum in a number of different ways, which are outlined in Fig. 3.3. It is, however, preferable not to use the terms *upfield* and *downfield* since these suggest that the magnetic field is varying, which is not the case for modern spectrometers which work at a fixed magnetic field and vary the frequency.

The ^{13}C NMR spectrum of PMe_3 exhibits a peak at 20.5 ppm. Should the peak for the methyl carbons in NMe_3 be at higher or lower frequency? Nitrogen is more electronegative than phosphorus and, therefore, will withdraw more electron density from the carbon in NMe_3 compared with the phosphorus in PMe_3. Thus, the carbon in the nitrogen-containing compound is more deshielded and so we expect a peak at higher frequency than for trimethylphosphine. The observed ^{13}C NMR peak for NMe_3 is at 47.5 ppm, at higher frequency as expected.

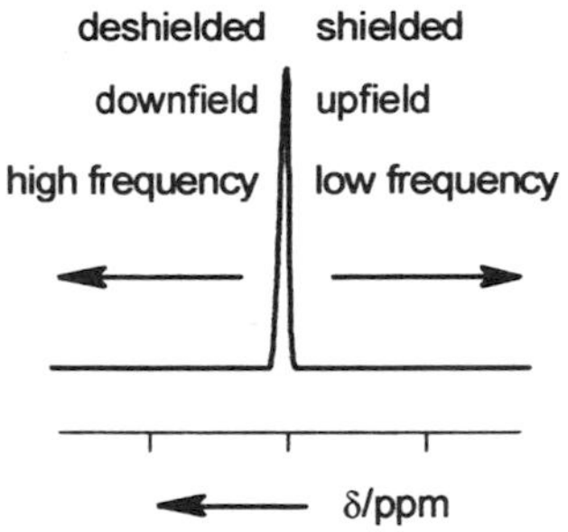

Fig. 3.3 Nomenclature used to describe the relative positions of two peaks in an NMR spectrum.

Chemical environment

Because the shielding of a nucleus is influenced by the type of atoms surrounding it and the bonding arrangements, the nuclei of a particular group in one molecule will resonate at a similar frequency in another molecule. Thus it is possible to use the chemical shift of a particular nucleus to determine the type of *chemical environment* around the nucleus under study. For the compound $Si(CH_2CH_3)_4$ there are two different chemical environments for the proton nuclei—the CH_3 and CH_2 protons. However, within each group the protons are equivalent. Similarly, there are two ^{13}C nuclei chemical environments (the CH_3 and CH_2 carbons) but obviously only one silicon chemical environment.

The range of chemical shift values that are observed depends on the nucleus involved. A relatively small range of +20 to –30 ppm is common for ^1H nuclei whilst larger ranges covering many hundreds of ppm chemical shift exist for heavier atoms. Figures 3.4 and 3.5 and Tables 3.2 and 3.3 give a few examples of typical chemical shifts for some of the more commonly encountered nuclei.

Question 3.2 How many different ^1H chemical environments are there in: (i) H_2O; (ii) $CH_3CH_2SCH_3$; (iii) $(CH_3)Pt(P(CH_3)_3)Cl_2$?

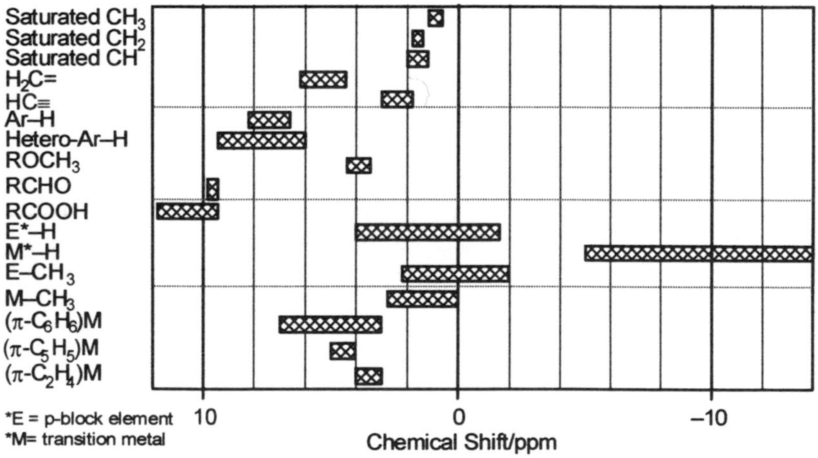

Fig. 3.4 ^1H chemical shift ranges for some proton environments.

	δ (ppm)		δ (ppm)
CH_3F	–226	$[BF_4]^-$	–150
C_2F_6	–91	PF_3	–36
C_6F_6	–163	SiF_4	–159
CF_3CO_2H	–78.5	XeF_2	–210
HF	–190	WF_6	164
F^-	125	MoF_6	277

Table 3.2 Representative ^{19}F NMR chemical shifts.

When compounds contain paramagnetic transition metals the above discussion on typical chemical shifts breaks down. In these cases peaks may not be observed at all, for reasons that are covered later, or their chemical shift values may be far from the expected values. As an extreme example the proton resonance for a π-bound C_6H_6 ligand in a compound such as $V(C_6H_6)_2$ should occur in the range 3–7 ppm, according to the data in Fig. 3.4, but because this compound contains V^0 which is a d^5 ion and hence is paramagnetic, the peak is actually observed nearer 300 ppm!

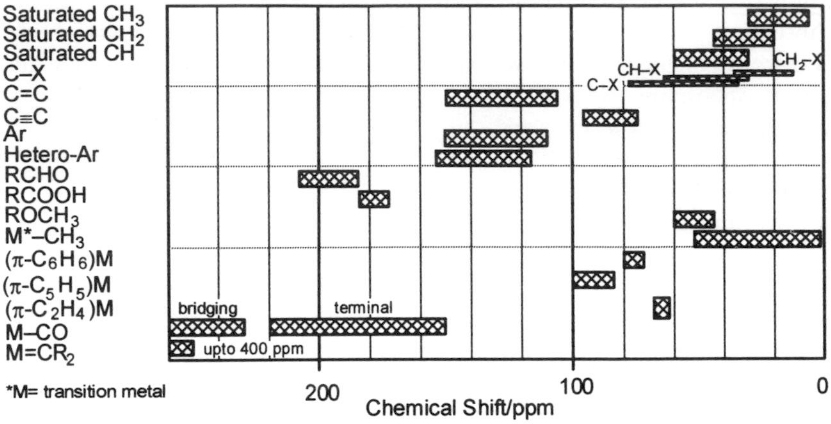

Fig. 3.5. ^{13}C chemical shift ranges for some carbon environments.

	δ (ppm)		δ (ppm)
PMe_3	–62	Me_3PO	36
PEt_3	–20	Et_3PO	48
PBu_3	–33	Bu_3PO	43
PPh_3	–7	Ph_3PO	29
$[Et_4P]^+$	40	$P(OMe)_3$	140
$[PF_6]^-$	–145	$P(OEt)_3$	138

Table 3.3 Representative ^{31}P NMR chemical shifts.

In many situations it is clear when nuclei are in different chemical environments. For example, in the complex $[(C_5H_5)_2RuH]^+$ the ^1H nuclei in the C_5H_5 ligand and the metal hydride are obviously chemically different and so separate peaks are observed in the proton NMR spectrum corresponding to these different environments. Conversely, in a highly symmetric molecule such as C_{60}, shown in Fig. 3.6, all the carbon atoms are equivalent and the ^{13}C NMR spectrum shows just a single peak. However, there are occasions when the distinctions are more subtle.

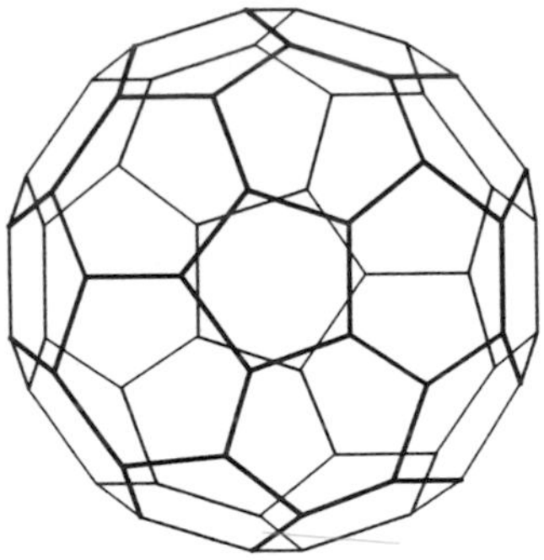

Fig. 3.6 The highly symmetric structure of C_{60}.

For example, consider the fluoride ligands of the IF_5 molecule, shown diagramatically in Fig. 3.7. The five fluoride ligands of this molecule are not all equivalent, four of them are in one plane and *trans* to a second fluoride, the fifth ligand is in a different plane and is *trans* to a lone-pair of electrons. It is not possible just by twisting and turning the molecule to swap the axial fluoride with any of the equatorial ligands, or vice versa. Thus there are two chemical environments for the fluorides, and we would expect to see two peaks in the ^{19}F NMR spectrum. In fact, we see two sets of peaks as shown in Fig. 3.8.

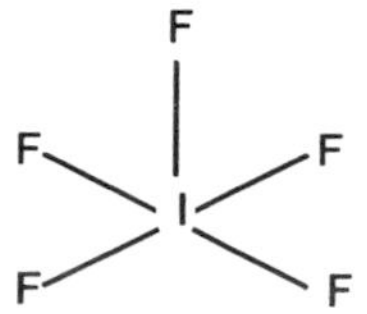

Fig. 3.7 The structure of IF_5.

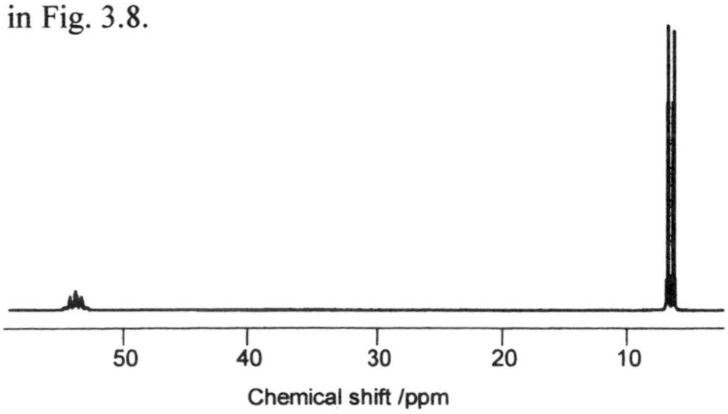

Fig. 3.8 The ^{19}F NMR spectrum of IF_5 in $CDCl_3$ solution.

The peaks around 7 ppm are due to the four fluoride ligands in the base of the pyramidal structure and the set of peaks around 55 ppm is due to the single axial fluoride. We will return to why the observed spectrum shows more peaks than we might initially expect very soon.

How many different phosphorus chemical environments are there in the *fac* and *mer* isotopes of a complex $[M(PR_3)_3Cl_3]$ shown in Fig. 3.9? For the *fac* form of the molecule all the phosphorus atoms are *trans* to a chloride ligand and therefore all are equivalent. For the *mer* form there are two phosphorus ligands *trans* to each other and the third one is *trans* to a chloride and therefore in a different chemical environment than the other two. Thus we could, in theory, determine whether a complex of the type $[M(PR_3)_3X_3]$ adopts the *fac* or *mer* conformers, just on the basis of the number of peaks observed in its ^{31}P NMR spectrum.

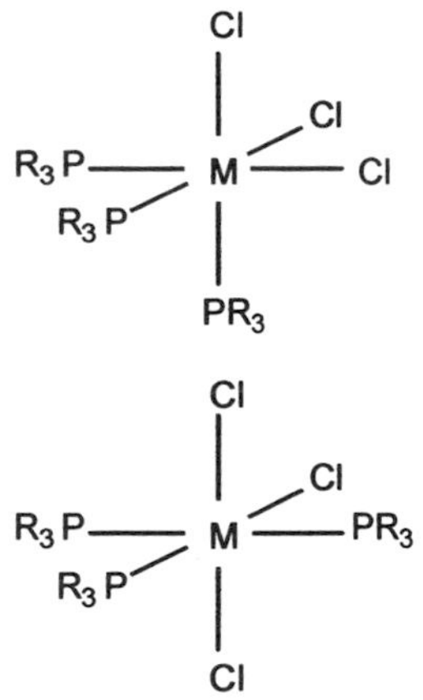

Fig. 3.9 *fac-* and *mer-*$[M(PR_3)_3Cl_3]$.

Integration

You will notice that in the ^{19}F spectrum of IF_5 shown in Fig. 3.8, the two peaks are of very different heights or, more correctly, the area under each set of peaks is different. This is due to the fact that each of the five fluoride nuclei under study will give rise to an equally intense NMR signal; however, four of these are equivalent and therefore their signals are superimposed to give a signal four times as intense as that due to the single axial fluoride.

For the *mer*-$[M(PR_3)_3Cl_3]$ complex, shown in Fig. 3.9, there are two different chemical environments, P–*trans*-P and P–*trans*-Cl, with two equivalent phosphorus atoms in the first environment and just one in the second environment so we should expect the ^{31}P NMR spectrum of this complex to exhibit two peaks in a 2:1 ratio.

Question 3.3 How many different ^{13}C chemical environments are there in $Mn(CO)_5Br$ and $(\eta^2\text{-}C_2H_4)Pt(PEt_3)_2$?

Thus the number of distinct peaks and their position in an NMR spectrum tells us the number and type of chemical environments that exist in a molecule for the nucleus under study and the relative intensity of each peak shows how many nuclei there are in each of these environments. However, it should be noted that it is not always possible to determine the number of nuclei in a particular environment based purely on intensities. This is most commonly encountered for ^{13}C spectra due to the varying rates at which these nuclei return to the ground state. Integration values may also be inaccurate for other nuclei, especially those that cover large chemical shift ranges.

Coupling

So far we have seen that NMR spectra can be recorded for a number of different magnetically active nuclei and that each set of nuclei in different chemical environments gives rise to separate peaks. There is, however, another important piece of information that is available from NMR spectra which helps further in spectral assignment. This arises from the fact that *non-equivalent* magnetically active nuclei *couple* to each other. That is, the small magnetic field arising from one nucleus can influence the magnetic field experienced by the other, non-equivalent, nuclei that are nearby in the molecule.

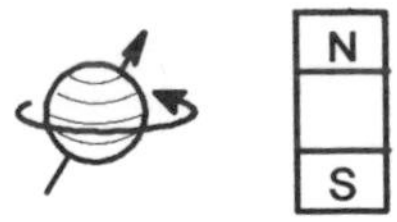

Fig. 3.10 Two different representations of a spin-½ nucleus.

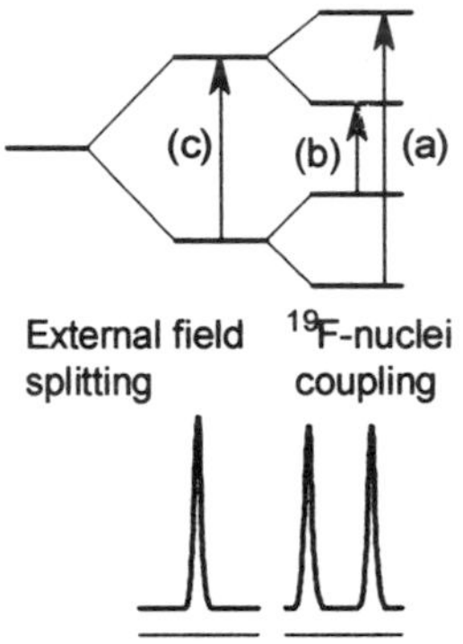

Fig. 3.11 Energy diagram showing the influence of the spin of the fluorine nucleus on the proton NMR transitions.

The fact that for I = ½ nuclei there are just two possible orientations results in a somewhat simplified but sometimes helpful analogy of thinking of a spin-½ nucleus as a small magnet, as shown in Fig. 3.10, which either aligns with or against the applied external magnetic field and thus increases or decreases the effective magnetic field experienced by other nuclei.

For example, the proton NMR spectrum of HF in solution consists of two equally intense peaks as does the ^{19}F NMR spectrum. There is only one proton and only one fluoride so it is unlikely that the two peaks arise from two different environments for each nucleus. Instead, because both the proton and the fluoride are magnetically active (I = ½), they each act as a small magnet and influence the magnetic field experienced by the other nucleus. Imagine each proton as a small bar magnet with two possible alignments either ↑ or ↓ corresponding to a spin of +½ or –½. The alignment of this small magnet can be either with the externally applied magnetic field or in the opposite sense. This either increases or decreases the effective magnetic field experienced by the nucleus under study. As we saw earlier according to the Boltzmann distribution both orientations are almost equally as likely since the energy difference between these states is small compared with the thermal energy, kT, which means that instead of one absorption being possible there are now two equally intense peaks.

Figure 3.11 shows what is happening in terms of an energy diagram. The applied magnetic field results in a relatively large splitting of the energy levels, labelled (c), onto which is superimposed the smaller splitting resulting from interaction with the other nucleus. The two possible transitions (a) and (b) between these levels result in the observation of a pair of peaks equally spaced either side of the position a single peak would have arisen if there had been no coupling (c).

The same effect will be seen in the ^{19}F NMR spectrum of HF, that is the two possible alignments of the proton will result in two transitions occurring of almost equal probability corresponding to the case when the magnetic moment of the proton (↑) increases the overall magnetic field experienced by the fluorine nucleus, and when it (↓) decreases it. So for both the ^{19}F and ^{1}H nuclei mutual coupling occurs resulting in a pair of peaks, called a doublet, in both the fluorine and proton NMR spectra, as shown in Fig. 3.12.

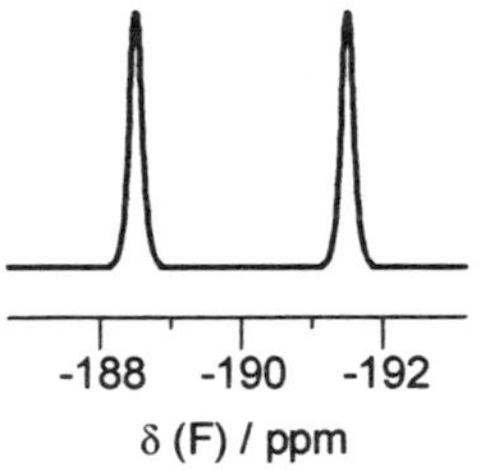

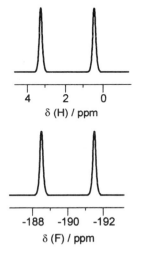

Fig. 3.12 Simulated (a) 200 MHz ^{1}H and (b) 188 MHz ^{19}F NMR spectra for HF plotted at 400 Hz cm^{-1}.

Because these two nuclei are coupling with each other, i.e. the coupling is mutual, the effect on the two nuclei is the same and therefore the two peaks of each set of doublets are equally separated when their positions are measured in Hertz. This separation is referred to as the *coupling constant* and can provide a number of pieces of information. Firstly, it allows us to be sure that the doublets in the ^{1}H and ^{19}F spectra arise from the same species, by virtue of both possessing the same coupling constant. Secondly, the magnitude of the coupling constant can provide some structural information as we will see later. Care needs to be exercised when measuring the coupling constant because it must to be measured in Hertz rather than ppm. If we do this for both the proton and the fluorine NMR spectra of HF and measure the frequency difference between the two peaks in each of the spectra then we do indeed see that the coupling constants are the same at around 530 Hz.

Determining the chemical shift value for a single peak is easy, but is slightly less obvious for a doublet or more complicated pattern, since the positions of the peaks in a multiplet are evenly distributed around a central point—the position if there was no coupling. This is taken as the chemical shift value. So for doublets the chemical shift value is the mid-point, measured on the ppm scale, and the coupling constant is the separation between the two peaks, measured in Hertz, as shown in Fig. 3.13.

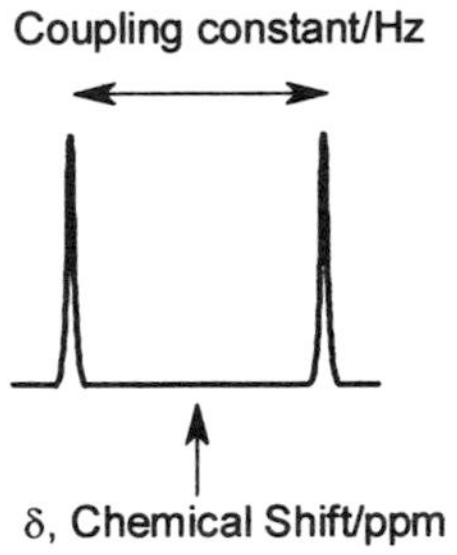

Fig. 3.13 Where to measure the coupling constant and chemical shift for a doublet.

If there are two chemically equivalent spin-active nuclei coupling to a single proton, such as in the HF_2^- ion, then there are four possible arrangements for the ^{19}F nuclei, i.e. ↑↑, ↑↓, ↓↑, or ↓↓, resulting in a net increase (↑↑) or decrease (↓↓) in the magnetic field for two of the arrangements, and for the other two (↓↑ or ↑↓) there will be no overall change in the effective magnetic field. So in one case the nuclei will resonate at a higher energy (or frequency) due to the additional magnetic field caused by the neighbouring nuclei aligning with the external field. When both these nuclei are aligned against the applied field a peak will be seen at a lower resonant frequency. The two mixed spin arrangements which result in no overall change in the magnetic field will not result in a change in frequency. We therefore would expect to see three equally spaced lines the intensities of which will be 1:2:1 to reflect the ratio of each arrangement as shown in Fig. 3.14.

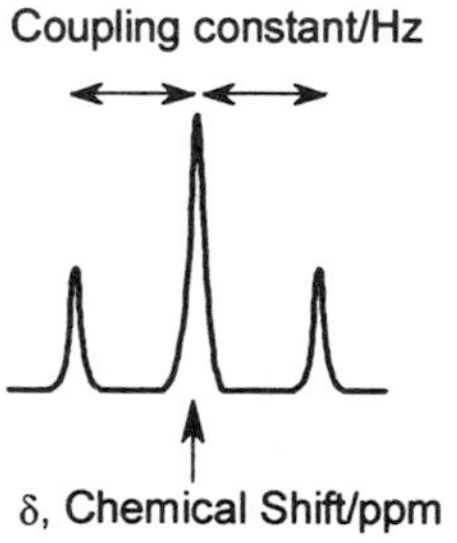

Fig. 3.14 The triplet spectrum expected for the HF_2^- ion.

It is common to see stick diagrams used to illustrate these patterns such as the examples shown in Fig. 3.15 for coupling to one, two, three, and four equivalent nuclei.

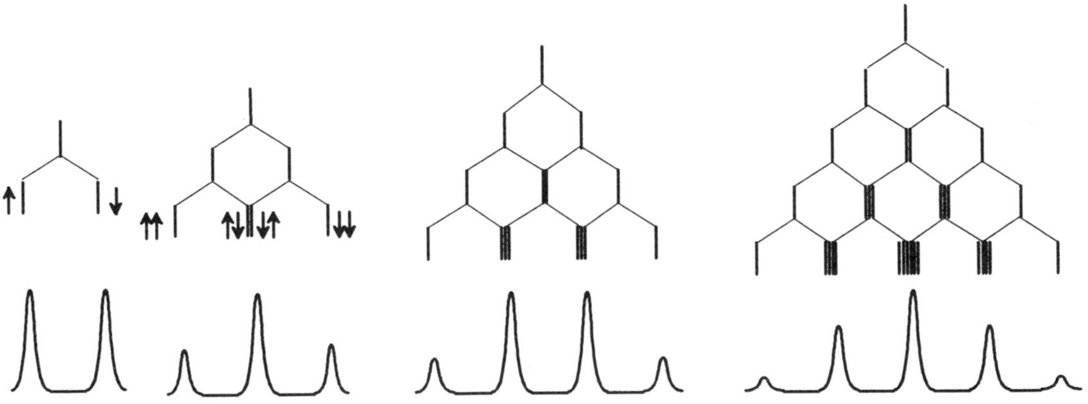

Fig. 3.15 Stick diagrams showing the coupling resulting from one, two, three, and four equivalent spin-½ nuclei.

In general, any nucleus will couple with every other magnetically active nuclei in the molecule provided they are chemically non-equivalent to give rise to a pattern consisting of

Question 3.4 The four peaks of a quartet recorded on a 200 MHz spectrometer are observed at 1.2, 1.4, 1.6, and 1.8 ppm. What is: (i) the chemical shift value; (ii) the coupling constant?

$$(2nI+1) \tag{3.7}$$

lines, where n is the number of coupling nuclei possessing a magnetic quantum number of I. For spin-½ nuclei this simplifies to $(n+1)$. Thus for the HF_2^- ion the two fluorine nuclei will couple with the single proton nucleus to give rise to two lines in the ^{19}F spectrum and the proton will couple to the two fluorine nuclei to produce a three-line pattern in the ^1H spectrum. The separation between each of the peaks within the pattern (measured in Hertz) is the coupling constant. The relative intensities of the lines can be calculated by hand as previously or for spin-½ nuclei are given by Pascal's triangle which is shown below.

	Number of equivalent coupling spin-½ nuclei	Name of pattern
1	0	singlet
1 1	1	doublet
1 2 1	2	triplet
1 3 3 1	3	quartet
1 4 6 4 1	4	quintet
1 5 10 10 5 1	5	sextet

The ^{31}P and ^{19}F NMR spectra of an ion of the type $[P_xF_y]^-$ are shown in Fig. 3.16. What is the simplest empirical formula for the anion?

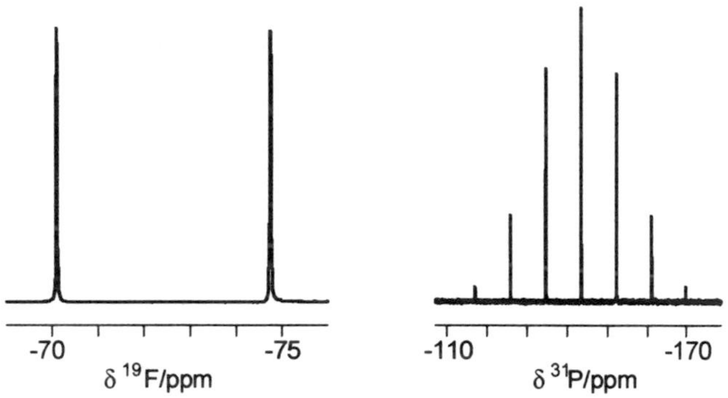

Fig. 3.16 ^{31}P and ^{19}F NMR spectra for an ion of the type $[P_xF_y]^-$.

Starting with the ^{19}F spectrum, this shows two peaks of approximately equal intensity—a doublet. This suggests that a set of one, or more, equivalent fluorine nuclei is coupling with one phosphorus atom (spin = ½) to yield a doublet ($2nI + 1 = 2\times1\times½ + 1 = 2$) in the NMR spectrum. The ^{31}P NMR spectrum exhibits a seven-line pattern in what looks like a binomial distribution. Since fluorine possesses a spin-½ nucleus a binomial seven-line pattern is most likely to arise from coupling to six equivalent fluorine nuclei ($2nI + 1 = 2\times6\times½ + 1 = 7$). So we have one phosphorus nucleus coupling to six equivalent fluorine nuclei to give a septet in the ^{31}P NMR spectrum and the six equivalent fluorine nuclei coupling with the single phosphorus nucleus resulting in a doublet in the fluorine spectrum. The simplest ion accounting for this is $[PF_6]^-$ with the fluorides being octahedrally arranged so that they are all equivalent. We can confirm that these two nuclei are coupling with each other by measuring the coupling constants in both spectra which should be the same—which they are in this case, at 709 Hz.

The coupling described above between two different nuclei is called *heteronuclear coupling*, but coupling can also arise from the same nuclei in different environments in which case it is referred to as *homonuclear coupling*.

Question 3.5 Predict the ^{31}P and ^{1}H NMR spectra for PH_3.

Figure 3.17 shows the 200 MHz ^{1}H NMR spectrum recorded for a sample of the tetrahedrally-based molecule $Si(OCH_2CH_3)_4$.

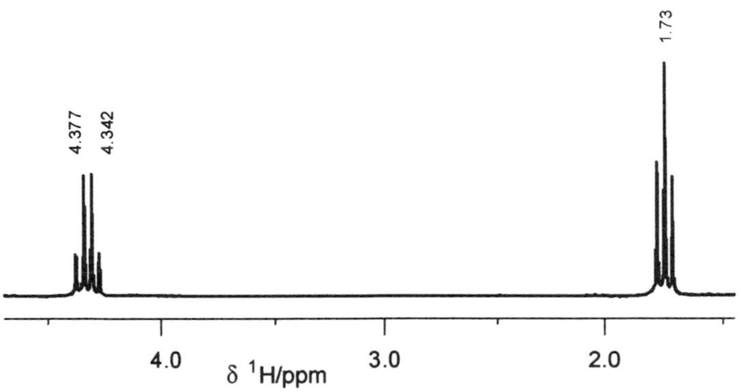

Fig. 3.17 The ^{1}H NMR spectrum of $Si(OCH_2CH_3)_4$ in $CDCl_3$.

Before attempting to assign this, or any, spectrum we need to determine the number of different chemical environments for the spin-active nuclei. The two CH_2 protons are chemically non-equivalent to the three CH_3 protons so there are two different chemical environments. We should therefore expect two sets of peaks. The spectrum does indeed show two chemical environments, a triplet signal at 1.73 ppm and a quartet centred at 4.35 ppm. Since the two sets of protons are chemically non-equivalent they will couple with each other. The CH_2 protons will couple with the three methyl protons resulting in a quartet and the three CH_3 protons will couple with the two CH_2 protons to yield a triplet. So the signal at 1.73 ppm must be due to the CH_3 protons and that at 4.35 ppm corresponds to the CH_2 protons. This relative ordering also fits with our idea of increasing chemical shift correlating with greater deshielding of the nucleus involved. The CH_2 protons are nearer the electronegative oxygen atom and therefore these protons will be more deshielded than the CH_3 protons.

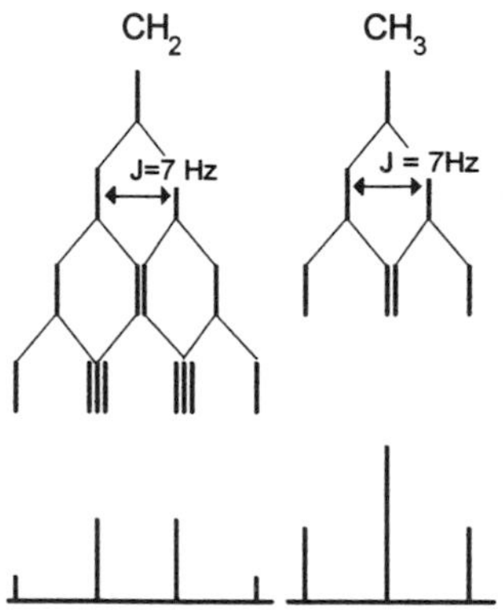

Fig. 3.18 Stick diagram showing how the couplings in $Si(OCH_2CH_3)_4$ arise.

The coupling constant, *J*, between these two nuclei can be determined by measuring the separation between any two peaks within a particular multiplet. The difference in ppm is converted to the Hertz scale by multiplying by the spectrometer operating frequency for the nucleus under investigation (in this case 200 MHz). So the coupling constant is

$$J = (4.377 - 4.342)\ \text{ppm} \times 200\ \text{MHz} = 7\ \text{Hz}$$

Figure 3.18 shows the stick diagram for this system which illustrates how the coupling patterns arise.

More about coupling constants

So far we have seen coupling between the nuclei of atoms that are directly bonded such as $[PF_6]^-$ and in $Si(OCH_2CH_3)_4$ between nuclei that are further apart. In fact coupling can extend over a considerable distance, and although it is most often propagated through areas of high electron density between atoms, i.e. bonds, it can also occur through space. It is usual when quoting coupling constants to include an indication of the number of bonds through which coupling has occurred. For example, the coupling between the fluorine and phosphorus atoms in $[PF_6]^-$ would be written as $^1J(PF)$ = 709 Hz where the superscript one refers to the single bond separating the coupling ^{31}P and ^{19}F nuclei. For $Si(OCH_2CH_3)_4$ we see coupling between the two non-equivalent sets of protons; this occurs through three bonds (H–C–C–H) and therefore would be written as $^3J(HH)$ = 7 Hz.

The magnitude of the coupling between any two nuclei for heteronuclear coupling depends on a variety of factors, including the nature of the nuclei and the bonding between them. The detailed mechanism that gives rise to this type of spin–spin coupling is beyond the scope of this book. One mechanism arises from contact between the two coupling nuclei by electrons in the bond between. Since only s-orbitals have a non-zero probability of the electron being at the nucleus, the coupling constant depends on the extent of s-orbital involvement in the bonding. For many nuclei there exists a relationship for ^1J coupling constants based on the percentage of *s*-orbital character of the bonding orbitals of the two nuclei of the form

$$J / \text{Hz} = k \times \%s(\text{A}) \times \%s(\text{X}) \tag{3.8}$$

where $\%s(\text{A})$ and $\%s(\text{X})$ are the percentage of *s*-character in the bonding orbital for nuclei A and X respectively and k is a proportionality constant. So $^1J(CH)$ will decrease as the hybridisation around the carbon changes from *sp* (i.e. 50% *s*–character) to sp^2 (33% *s*–character) to sp^3 (25% *s*–character). Similarly, as the distance between two coupling nuclei increases the coupling weakens, so ^1J coupling constants are usually larger than ^2J, and so on.

Coupling constant	Magnitude (Hz)
$^2J(^1H,^{13}C)$ in C_2H_6	125
$^2J(^1H,^{13}C)$ in C_2H_4	160
$^2J(^1H,^{13}C)$ in C_2H_2	250
$^1J(^{19}F,^{31}P)$	500–1500
$^2J(^{19}F,^1H)$	10–50
Terminal $^1J(^{11}B,^1H)$	100–200
Bridging $^1J(^{11}B,^1H)$	<100
$^1J(^{31}P^{V},^1H)$	400–1100
$^1J(^{31}P^{III},^1H)$	100–250
$^2J(^1H,^1H)$ on sp^3-C	10–20
$^2J(^1H,^1H)$ on sp^2-C	2–10

Table 3.4 Some typical coupling constants.

There are many factors influencing the coupling constants between two nuclei including the magnitude of the magnetogyric ratios of the two nuclei, the distance and bonding between them, and their angular arrangement. Although it is not easy to predict these values a few of the more typical ranges for some of the common nuclei are listed in Table 3.4.

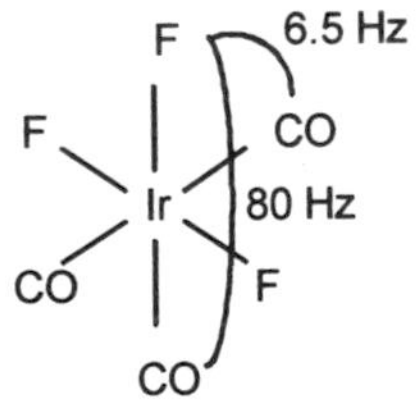

Fig. 3.19 Observed $^2J(^{13}C^{19}F)$ coupling constants in the complex *mer*-$[Ir(CO)_3F_3]$.

One commonly observed effect which can be useful in assigning coupling constants in NMR spectroscopy is that *trans* coupling constants are often considerably larger than *cis* ones. For example, Fig. 3.19 shows an example where the *trans* $^2J(^{13}C^{19}F)$ coupling constant is more than ten times larger than the corresponding *cis* one. A simple view of why this might be is that the two coupling atoms, if *trans* to each other, can interact with opposite parts of the *same* orbital thus providing a more direct mechanism for the coupling interaction than in the *cis* arrangement.

Labelling spin-systems

There is a shorthand nomenclature used to identify the type and number of spin-active nuclei in a molecule, which give rise to similar patterns. When the coupling constants are much smaller than the difference in chemical shifts between sets of peaks we observe patterns which are easy to interpret and understand, for example doublets, triplets, etc. These are called *first-order systems*. The shorthand notation uses letters that are near together in the alphabet to describe magnetically active nuclei that resonate at similar frequencies and letters that are far apart for those that resonate at very different frequencies. So for example the proton spectrum of the compound $Cl_3Si(OCH_2CH_3)$ will give rise to a triplet and a quartet. Both nuclei resonate at fairly similar frequencies, they are only a few ppm apart, so letters near to each other in the alphabet are used. If we ignore the other low abundance nuclei then we can describe this as an A_2B_3 system. The same A_2B_3 system exists in, for example, CF_3CF_2Cl or trigonal bipyramidal $M(PR_3)_5$ complexes and so the NMR spectra of all these compounds should display a triplet and a quartet. In IF_5 there are two different fluoride environments in the ratio 1:4 thus we describe this as an AB_4 spin-system.

If the nuclei are different or their resonance frequencies are very well separated then we would describe the system using letters that are far apart in the alphabet, e.g. HF is described as an AX spin-system. Likewise we could describe the compounds PF_2Cl and PH_2I as AX_2 spin-systems and their NMR spectra would show similar characteristic patterns.

Question 3.6 What would the typical splitting pattern be for the A and X nuclei of (i) AX, (ii) AX_2, and (iii) AX_3 systems?

Magnetic versus chemical equivalence

So far we have referred to nuclei as being chemically equivalent or non-equivalent and used this as the basis of interpreting their NMR spectra. However, there is another type of equivalence that needs to be considered. For example, the dianion $[H_2P_2O_5]^{2-}$ shown in Fig. 3.20 appears, at first sight, to possess just one proton and one phosphorus environment. However, although it is true that the pairs of atoms H_a and H_b and P_a and P_b are chemically equivalent, they differ in a more subtle way.

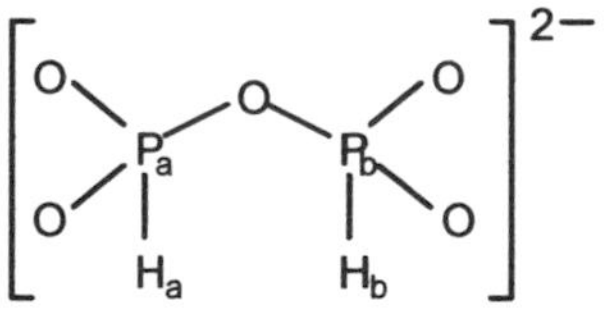

Fig. 3.20 The dianion $[H_2P_2O_5]^{2-}$.

Consider the interactions of P_a with the other magnetically active nuclei in the molecule. Firstly, P_a will couple to P_b in the same way that P_b couples to P_a. But the coupling of P_a to H_a and H_b will be different, one of them is directly bound to P_a while the other is separated by three bonds. In a similar way H_b will couple differently to P_a and P_b. The two phosphorus nuclei are chemically equivalent but because they couple differently with sets of other chemically equivalent nuclei (H_a and H_b) they are termed *magnetically non-equivalent*. Such a system is labelled as AA'XX' to distinguish between the chemically equivalent but magnetically non-equivalent sets of nuclei.

Question 3.7 What are the spin-system labels for (i) $[PF_6]^-$, (ii) PCl_2Et, and (iii) $H_2C{=}CF_2$?

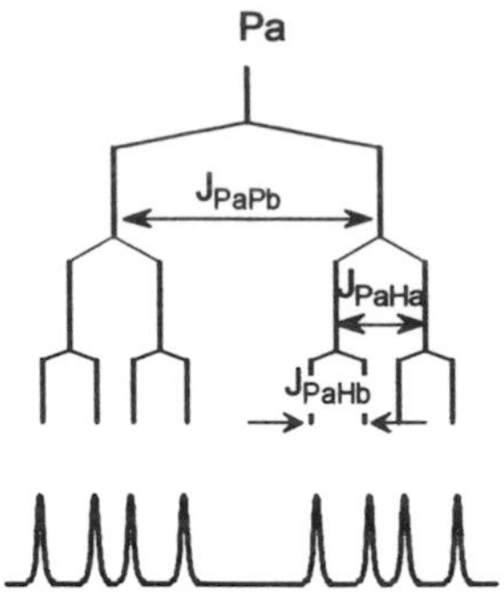

Fig 3.21 Expected ^{31}P NMR spectrum for the AA'XX' system $[H_2P_2O_5]^{2-}$.

Because the nuclei are magnetically non-equivalent they may now couple with each other. So the phosphorus NMR spectrum for nucleus P_a is expected to be a doublet (due to coupling with P_b) of doublets (H_a) of doublets (H_b). The pattern of the ^{31}P NMR signal due to P_b will be the same and at the same chemical shift position because the two chemical environments of P_a and P_b are identical. This pattern, and its derivation, is shown in Fig. 3.21. In a similar way the proton NMR spectrum is predicted to consist of a doublet of doublets of doublets.

Question 3.8 What would be the appearance of the ^{31}P NMR of the $[H_2P_2O_5]^{2-}$ ion if (i) $J(P_aH_b)=0$, (ii) $J(P_aP_b)=J(P_aH_a)$?

Decoupling

There are occasions when, because of the large number of potential coupling interactions either due to chemical or magnetic non-equivalence, the interpretation of an NMR spectrum may be very difficult. Methods of recording spectra exist such that the coupling between certain nuclei can be removed. These experiments are performed most often for systems which contain a lot of protons, such as when recording ^{13}C and ^{31}P NMR spectra. This is achieved by providing energy over a range of frequencies that cover the proton resonances to saturate any transitions that are causing the coupling. It is normally denoted using a formalism such as $^{13}C\{^1H\}$ which means the carbon-13 spectrum recorded with proton decoupling, that is coupling between ^{13}C and ^1H nuclei has been removed. As an example consider the σ-bound C_6H_5 ligand shown in Fig. 3.22. This has a number of different carbon and proton environments, but we will concentrate on the patterns expected for the carbon labelled as C-3. In the absence of proton decoupling there would be coupling between the C-3 carbon and the proton attached to it (H-3) resulting in a doublet, but coupling can also occur to the two protons H-2 and H-4 so that each of the doublet peaks becomes a triplet. Further coupling can occur with the protons H-1 and H-5 resulting in further splitting of every line in the doublet of triplets into further triplets. It should be obvious that potentially the spectrum will consist of a very large number of weak peaks arising from the carbon nuclei in the ligand. The removal of this coupling results in a single carbon environment giving rise to a single peak, as shown in the spectrum in Fig. 3.22.

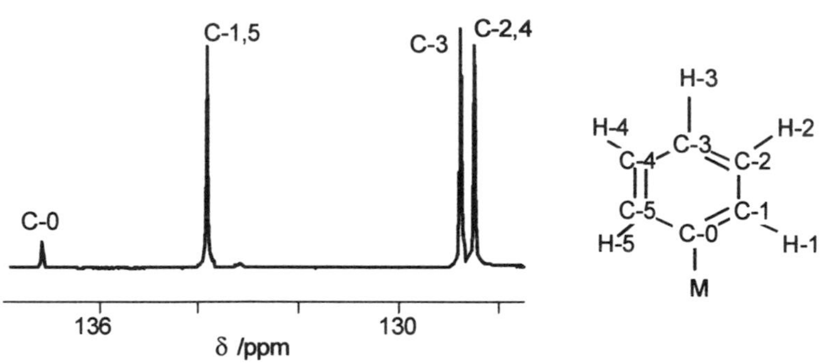

Fig. 3.22 A σ-bound C_6H_5 ligand and its $^{13}C\{^1H\}$ NMR spectrum.

Spin-dilute systems

Most of the spin-active nuclei considered so far have been 100% abundant. However, there are many elements which possess a number of isotopes only some of which are spin active. Such systems which possess less than 100%

abundant spin-active nuclei are often described as *spin dilute*. One such nucleus is ^{13}C which occurs at 1.1%. In the previous example no homonuclear coupling between different ^{13}C nuclei is observed; this is because of the low natural abundance of ^{13}C. There is just over a 1% chance of any carbon being carbon-13, the chance that a second atom is also carbon-13 is around 0.01% (i.e. 1.1% × 1.1%) which is so low as not to be practically observed.

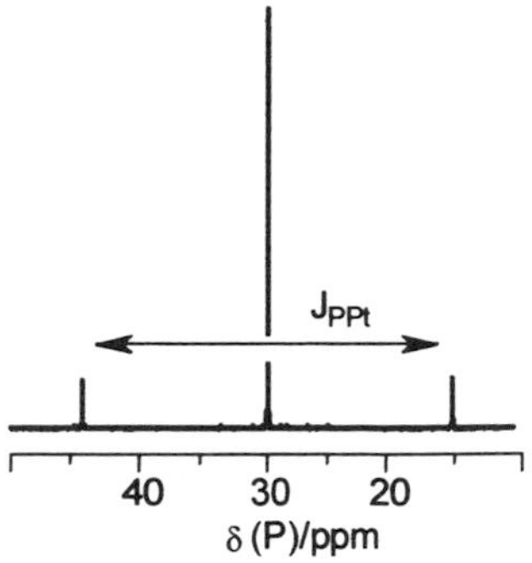

Fig 3.23 The $^{31}P\{^1H\}$ NMR spectrum of *cis*-$[Pt(PEt_3)_2Cl_2]$.

The $^{31}P\{^1H\}$ NMR spectrum of the complex *cis*-$[Pt(PEt_3)_2Cl_2]$ is shown in Fig. 3.23 and the structure of the compound is shown in Fig. 3.24. The two phosphorus nuclei are chemically equivalent and should therefore give rise to just one signal. However, 33.8% of the platinum atoms are spin active ($I = \frac{1}{2}$) and will couple, whilst the remaining 66.2% are not. The NMR spectrum of those molecules that contain a spin-active platinum atom will result in a doublet separated by ^1J(Pt–P) while the others, which do not contain a spin-active platinum nucleus, will result in a single peak. The observed spectrum is the sum of these two superimposed parts.

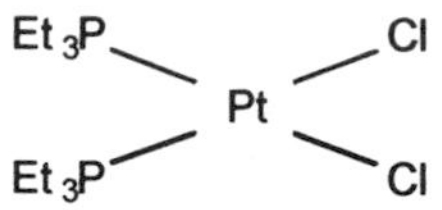

Fig. 3.24 The structure of the complex *cis*-$[Pt(PEt_3)_2Cl_2]$.

The relative intensities of the peaks can be explained fairly easily—the central peak must correspond to the molecules which do not contain spin-active platinum (*ca.* 66%) and the other two peaks together will account for the remaining 34% of the molecules and will thus each be (34/2)% of the total intensity. These weaker peaks are often referred to as *satellites* to the main peak. Of course we can also record the spectrum of the spin-dilute nucleus, in which case we only observe a spectrum from molecules containing the spin-active nucleus and therefore coupling will be observed with the other spin-active nuclei. So the ^{195}Pt NMR spectrum of *cis*-$[Pt(PEt_3)_2Cl_2]$ will consist of a triplet due to coupling with the two equivalent phosphorus atoms, the coupling constants due to the other low-abundance spin-active nuclei being too small to be observed.

A very common case of satellite structure is often seen in proton spectra of organic compounds. Since 1.1% of carbon is carbon-13 which is spin active ($I = \frac{1}{2}$), coupling to the proton nuclei will occur resulting in weak satellite peaks (0.55% the total intensity) either side of the main peak.

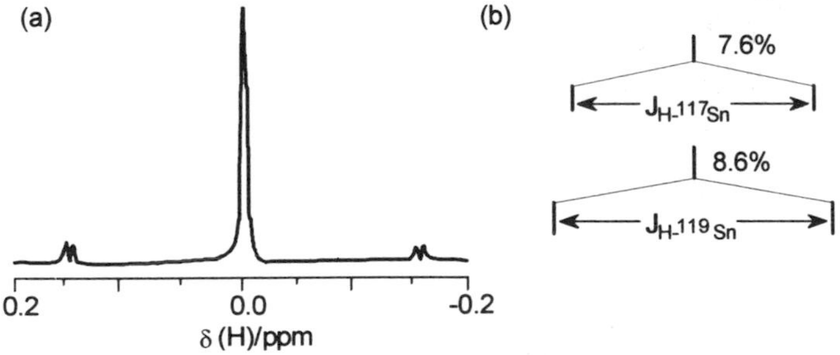

Fig. 3.25 (a) The ^1H NMR spectrum of $Sn(CH_3)_4$ and (b) stick diagrams showing the derivation of the peaks.

The ^1H NMR spectrum of $Sn(CH_3)_4$ shown in Fig. 3.25(a) consists of a singlet arising from the 12 equivalent protons. However, tin has two spin-active nuclei, ^{117}Sn (7.6%, $I = \frac{1}{2}$) and ^{119}Sn (8.6%, $I = \frac{1}{2}$), so molecules which contain either of these spin-active nuclei will couple with the protons to yield two sets of doublets equally spaced about the central peak. The splitting patterns are shown in Fig. 3.25(b).

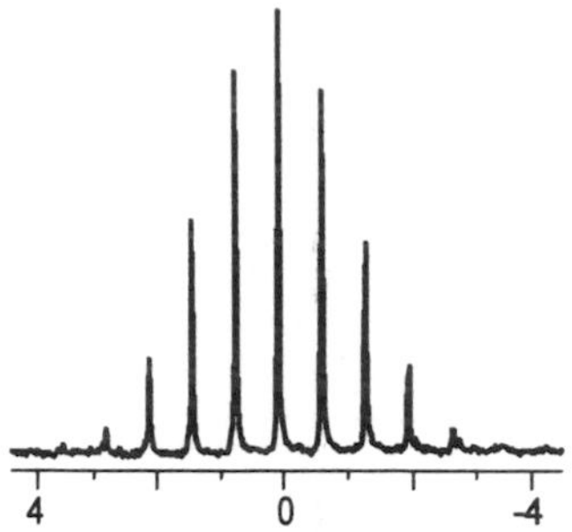

Fig. 3.26 The ^{119}Sn NMR spectrum of $Sn(CH_3)_4$.

Question 3.9 Predict the $^{13}C\{^1H\}$ NMR spectrum of $SnMe_4$

What should the ^{119}Sn spectrum look like? The tin atoms can either be 117-tin or 119-tin, but never both, so there will be no ^{117}Sn satellites on the ^{119}Sn signal. However, coupling to the proton nuclei will occur resulting in a binomial pattern consisting of 13 lines ($2nI + 1 = 2\times12\times\frac{1}{2} + 1 = 13$). The spectrum is shown in Fig. 3.26. However, the very weakest peaks of the pattern (1:12:66:220:495:792:924:792:495:220:66:12:1) are not visible.

Other elements with less than 100% abundant spin-active nuclei which are fairly common in inorganic chemistry include B, Si, Se, Sn, W, Pt, Cd, and Hg.

Non-spin-½ systems

So far only spin $I = \frac{1}{2}$ nuclei have been considered, but there are a number of common nuclei which have greater magnetic moments. By virtue of the larger magnetic quantum number these nuclei also possess a *quadrupole moment*. Although, in principal, the interpretation of spectra for nuclei with $I > \frac{1}{2}$ is no more difficult than for spin-½ nuclei, the presence of a quadrupole can result in broad peaks, which makes their application more limited. The factors that give rise to this broadening are covered in more detail in a later section. Table 3.5 lists data for some of the more commonly studied quadrupolar nuclei.

Nucleus	Spin	Natural abundance (%)	Relative NMR frequency (MHz) (B_0 = 4.7 T)	Magnetogyric ratio (10^7 T^{-1} s^{-1})	Receptivity relative to ^{1}H (1.00)	Standard reference compound	Common range (ppm)
^{7}Li	3/2	92.6	77.7	10.39	0.27	Li^+(aq)	−10 – +5
^{10}B	3	19.6	20.7	2.87	4×10^{-3}	$Et_2O.BF_3$	−150 – +100
^{11}B	3/2	80.4	64.2	8.58	0.13	$Et_2O.BF_3$	−150 – +100
^{14}N	1	4.7	14.4	1.93	1×10^{-3}	$MeNO_2$	−400 – +550
^{27}Al	5/2	33.8	52.1	6.97	0.22	$[Al(H_2O)_6]^{3+}$	−250 – +250

Table 3.5 NMR properties of some of the more commonly studied quadrupolar nuclei.

As outlined in eqn 3.1, when a spin-active nucleus is placed in a magnetic field a number of quantized energy levels will arise depending on the magnetic quantum number. So, for example, for ^{11}B which has $I = \frac{3}{2}$ the following states will be generated: $m = \frac{3}{2}, \frac{1}{2}, -\frac{1}{2}$, and $-\frac{3}{2}$. Since the only transitions allowed are those which have $\Delta m = \pm1$, three transitions are possible, $\frac{3}{2} \leftrightarrow \frac{1}{2}$, $\frac{1}{2} \leftrightarrow -\frac{1}{2}$, and $-\frac{1}{2} \leftrightarrow -\frac{3}{2}$, and since all of these occur at the same energy a single peak will be observed for a single non-coupling ^{11}B nucleus. However, the fact that there are four allowed nuclear spin orientations rather than just two does make a difference to the patterns observed when coupling to other nuclei occurs.

The BH_4^- anion adopts a tetrahedral structure, so all the protons are chemically and magnetically equivalent. The central boron atom may be either ^{10}B or ^{11}B and both will couple with the four protons. In the case of the central atom being ^{11}B (80% probability, $I = \frac{3}{2}$) the protons will couple so that the ^{1}H NMR spectrum will consist of a pattern of four lines ($2nI + 1 = 2\times1\times\frac{3}{2} + 1 = 4$). Since each of the four possible values of m are equally likely for the boron nucleus and each will affect the magnetic field experienced by the proton, four equal intensity lines are observed in the ^1H NMR spectrum. So we would expect to see a 1:1:1:1 quartet in the proton NMR spectrum of $[^{11}BH_4]^-$, as shown in Fig. 3.27(a).

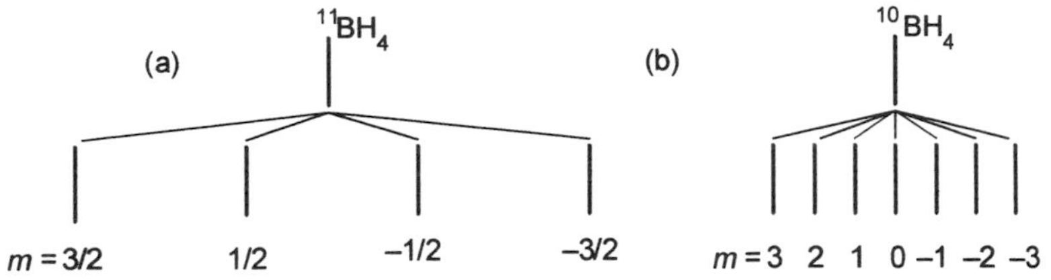

Fig. 3.27 The splitting patterns expected for the ^1H NMR of the (a) $^{11}BH_4^-$ and (b) $^{10}BH_4^-$ ions.

However, this is not the end of the problem because boron has a second isotope which is also magnetically active: ^{10}B, I = 3, 20% abundant. By applying the $(2nI + 1)$ formula we should expect the proton spectrum for the $^{10}BH_4^-$ ions to exhibit a seven-line pattern as is shown in Fig 3.27(b). The actual spectrum will be a combination of these two patterns. Because there will be negligible difference between the chemical shifts of the protons in the $^{10}BH_4^-$ and $^{11}BH_4^-$ ions the two patterns will appear to be centred around the same point. However, we might expect the coupling constants to differ since the magnetogyric ratios for the two nuclei are different. The relative intensities of the lines in these overlapping multiplet patterns can be predicted based on the natural percentage abundance of the nuclei involved and the number of lines within each pattern they are responsible for. For the ^{11}B-containing species, which represent 80% of all borohydride ions, four lines are generated, so each is 20% of the total intensity. Similarly, for the ^{10}B-containing ions (20% natural abundance) seven lines are generated, so each line is a little less than 3% of the total intensity. Figure 3.28 shows the observed proton NMR spectrum, and the peaks due to ^{10}B-containing species are marked with a plus sign and those for $^{11}BH_4^-$ with an asterisk.

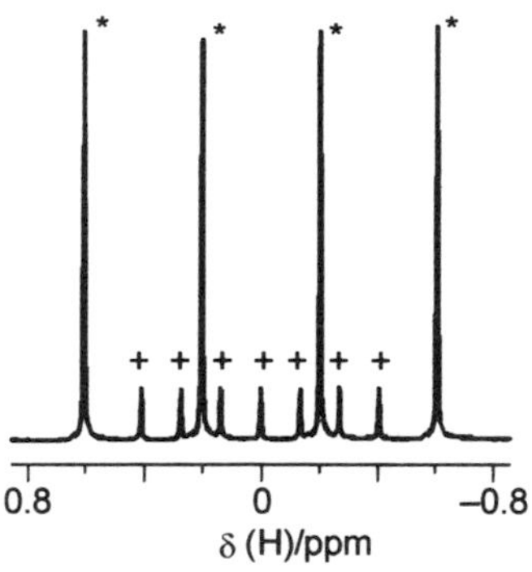

Fig. 3.28 The observed proton NMR spectrum for $K^+BH_4^-$ dissolved in D_2O.

We could also record either the ^{10}B or ^{11}B NMR spectrum; however, because of the higher natural isotopic abundance and receptivity, it is more usual to study ^{11}B NMR. The ^{11}B NMR spectrum of $^{11}BH_4^-$ will consist of one boron resonance (since there is only one boron environment) coupled to four equivalent spin-½ protons. This will result in a five-line pattern ($2nI + 1 = 2\times4\times\frac{1}{2} + 1 = 5$) and in this case because the splitting is due to spin-½ nuclei the pattern will be binomial in appearance and the relative intensities of the lines given by Pascal's triangle so a 1:4:6:4:1 quintet should be seen.

Of course molecules containing ^{10}B will not show in an ^{11}B NMR spectrum. However, if we were to record the ^{10}B NMR spectrum of the 20% of molecules that do contain $^{10}BH_4^-$ then again the single boron resonance would be split due to coupling to four equivalent protons resulting in a 1:4:6:4:1 quintet pattern.

Exchange processes

Earlier we saw how additional lines may arise in an NMR spectrum due to magnetic non-equivalence of nuclei. There is also the possibility that fewer lines than expected may be seen. This is quite common within the field of organometallic chemistry.

For example, the complex [FeI(CO)$_2$Cp] adopts the structure shown in Fig. 3.29. It would be expected that the protons on the cyclopentadienyl ring sitting over the carbonyl ligands will be different from those sitting over the iodide ligand. However, the proton NMR spectrum, Fig. 3.30(a), shows just a single peak with no coupling. This is because the cyclopentadienyl ring is rotating around its central axis which means that during the lifetime of the NMR experiment the protons experience all possible environments giving rise to a single peak with an average chemical shift value. This form of motion also removes the potential for coupling between the protons because they all become equivalent. Similarly, the ^{13}C NMR spectrum shown in Fig. 3.30(b) also exhibits just a single intense peak in the aromatic region.

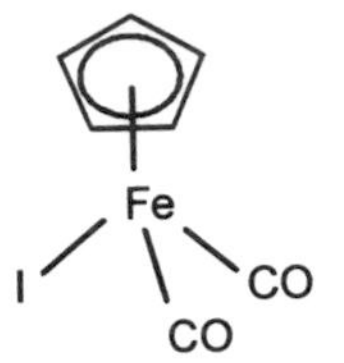

Fig. 3.29 The structure of the complex [FeI(CO)$_2$Cp].

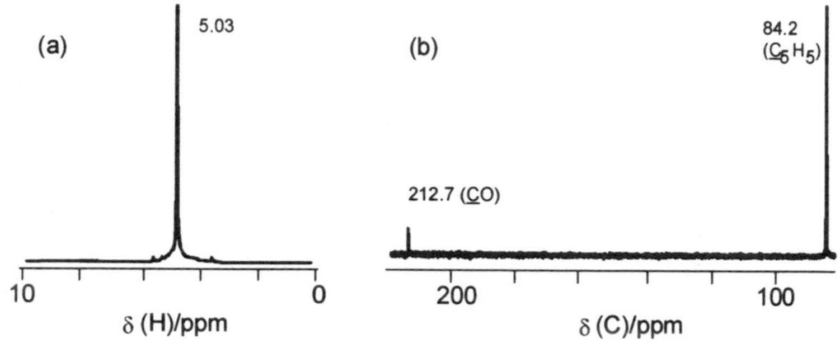

Fig. 3.30 The (a)^{1}H and (b) ^{13}C NMR spectra of the complex [FeI(CO)$_2$Cp].

In general, if τ is the lifetime of a molecule in a particular conformation, then its energy will be uncertain according to $\Delta E \approx h/2\pi\tau$. When this energy corresponds to the difference between resonance frequencies ($2\pi\Delta\nu$) of the two states a single broad peak is observed. If the lifetime is very much longer than the equivalent frequency separation the uncertainty is smaller and so two separate resonances are observed. If, on the other hand, τ is very much less than $2\pi\Delta\nu$ then a sharp singlet will be observed corresponding to an average chemical environment. The expected spectra are shown in Fig. 3.31.

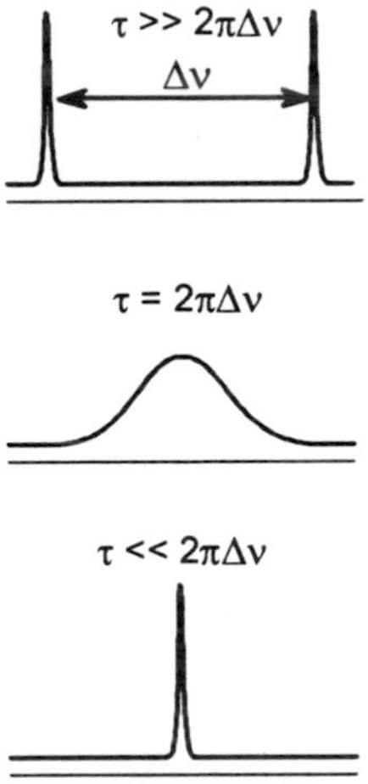

Fig. 3.31 The effect of the lifetime of an excited state on the observed spectra.

These effects are observed in a number of transition-metal systems. One representative example is the ^{1}H NMR spectrum of the complex [Rh(C$_5$H$_5$)(C$_2$H$_4$)$_2$], Fig. 3.32. At low temperatures four peaks are observed for the protons of the C_2H_4 ligands. This is because the inner (H_a) and outer (H_b) protons are in slightly different environments. However, as the temperature is raised the energy required to break the π-bond between the metal and alkene is overcome and the alkene undergoes a propeller-type of rotation around the metal–alkene σ-bond. This results in an exchange of the protons from the formerly non-equivalent environments, as shown in Fig. 3.33, and results in a broadening of the NMR signals. As room temperature is reached the speed of rotation increases until eventually it is rapid enough that only a single average chemical environment is experienced by the C_2H_4

protons, and a single peak is observed. A single peak is also observed for protons of the cyclopentadienyl ligand because this too is rotating.

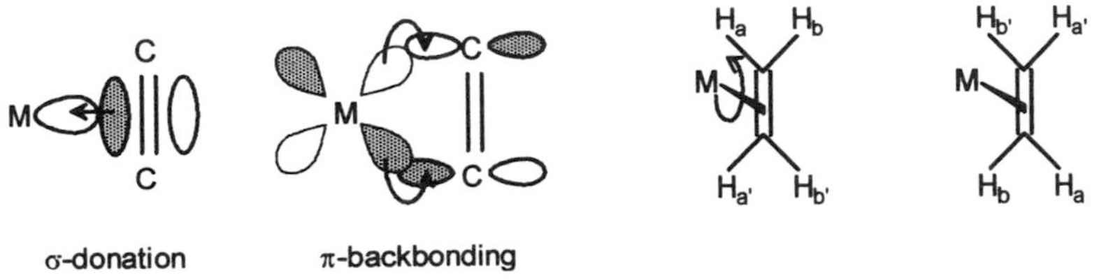

Fig. 3.33 The metal–alkene bonding scheme that allows propeller-like rotation of an alkene.

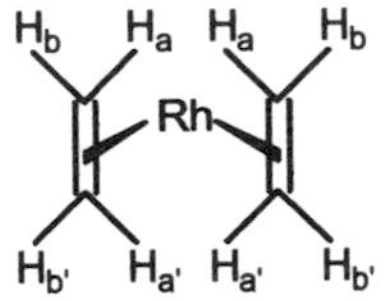

Fig. 3.32 The AA'BB' spin system of $RhCp(C_2H_4)_2$; for clarity the Cp ligand is not shown.

It is fairly common for proton environments on a number of π-bound organometallic systems to exchange near room temperature. This may arise from rotation of the ligand, as detailed above, or due to changes in the mode of bonding. For example, frequently the four protons attached to the terminal carbon atoms in the π-bound allyl ligand C_3H_5 appear as a single resonance because of exchange. The mechanism believed to be responsible for this is a change in bonding mode of the allyl ligand from being π-bound to σ-coordinated followed by a rotation of one end of the ligand around the central C–C bond before the π-bonding mode is reattained. The effect of this is to swap protons that were on the underside of the allyl ligand with those on the top, as shown in Fig. 3.34. There is equal probability that the σ-bond will be formed at either terminal carbon atom and so all four of the protons H_a, H_b, H_d, and H_e will exchange.

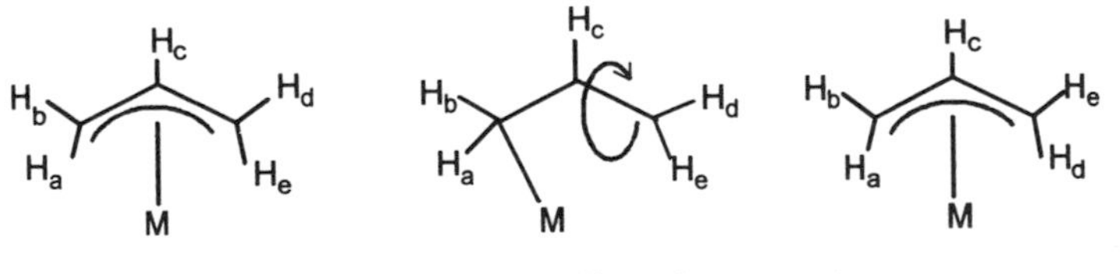

Fig. 3.34 The mechanism by which terminal allyl protons become equivalent.

Of course this mechanism does not allow the central proton to swap with the terminal ones and therefore a separate signal is observed for this, usually appearing as a quintet due to coupling with the other four protons that are equivalent due to the exchange process.

Another common case when exchange occurs is for many trigonal bipyramidal compounds which frequently, at room temperature, exhibit just a single resonance for all the ligands. For example, complexes of the type $[Rh(PR_3)_5]^+$ formally have two different phosphorus environments arising from the axial and equatorial ligands. At low temperature the ^{31}P NMR spectrum consists of a doublet of quartets and a doublet of triplets. This is due to coupling between the two different phosphorus environments and with the spin-½ rhodium nucleus, as shown in Fig. 3.36(a). At room temperature rapid motion occurs, such as that proposed by Berry shown in

Fig. 3.35. This allows the axial and equatorial ligands to swap and so become equivalent resulting in the rhodium atom 'seeing' five equivalent phosphorus nuclei. The resulting ^{31}P spectrum consists of a doublet at the weighted-average chemical shift value as shown in Fig. 3.36(b). The rhodium–phosphorus coupling constant also becomes the weighted average of the two static coupling constants.

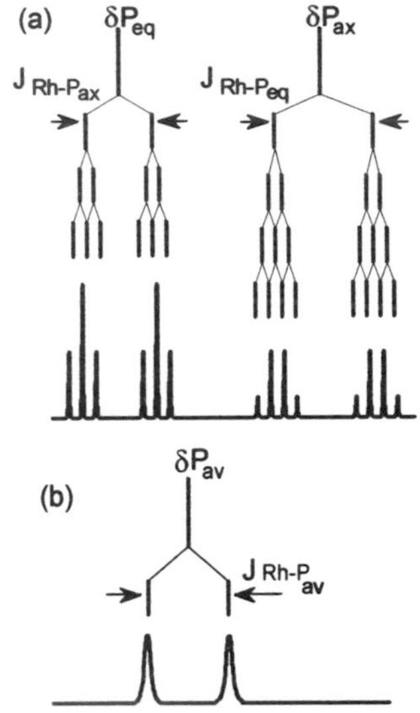

Fig. 3.36 The predicted NMR patterns for a complex of the type $[Rh(PR_3)_5]^+$ at (a) low temperature and (b) room temperature.

$$J(\mathrm{Rh-P})_{\mathrm{av}} = \frac{2 \times J(\mathrm{Rh-P_{ax}}) + 3 \times J(\mathrm{Rh-P_{eq}})}{5}$$

$$= \frac{2 \times 140 + 3 \times 210}{5} = 182\ \mathrm{Hz}$$

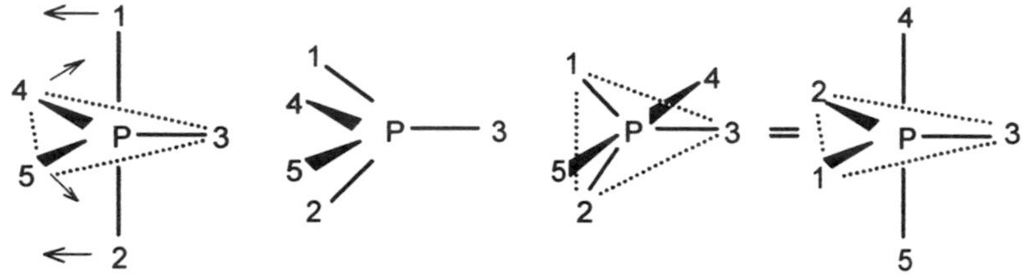

Fig. 3.35 The Berry mechanism that swaps axial and equatorial ligands in trigonal bipyramidal complexes.

These processes are all examples of *intramolecular* exchange, that is the ligands remain coordinated, but they experience an average environment due to thermal motion. *Intermolecular* exchange processes are also possible, in this case nuclei exchange between two or more different molecules. For example, proton resonances for the –OH and –NH groups are often broad or even absent due to chemical intermolecular exchange with the solvent. If the chemical exchange is slow then the signal will still be well resolved; however, as the process becomes more rapid the peak will become broader. Once the exchange process becomes sufficiently fast that the nucleus experiences an average chemical environment, a peak will be observed at an average chemical shift value. The fact that exchange is occurring between different molecules means that intramolecular coupling is lost.

Question 3.10 At low temperature two peaks are observed in the ^{13}C NMR spectrum of $Co_2(CO)_8$ at 182 and 234 ppm. Assign these peaks. If these two peaks arise from 6 and 2 carbon environments, respectively, at what position would you expect to observe a single peak due to thermal averaging?

Relaxation processes

As with all forms of spectroscopy there has to be some process or processes by which the perturbed energy states may return to their equilibrium state. If this did not happen then the two different energy states would become equally populated resulting in loss of the signal. It has already been noted that the population difference between nuclear spin-states is very small and therefore a method for reattaining the ground state is important in NMR spectroscopy. There are two main methods by which this may occur, these are known as *spin–lattice* and *spin–spin* relaxation. Typically for liquid-phase NMR spectroscopy both these relaxation times are of the order of a second under most circumstances.

Although there is not sufficient space to go into this subject in a rigorous manner here, both these relaxation mechanisms have an important bearing on the observed spectrum. Spin–lattice relaxation processes are those in which the energy that caused the nucleus to flip spin-state is given up to its surroundings, for example by collision with other molecules in solution.

If the NMR solution is very viscous then the natural Brownian motion is reduced and so the rate at which spin–lattice relaxation occurs is reduced. For example, the ^{19}F spectrum of neat, viscous IF_5 is shown in Fig. 3.37. Comparison with that shown in Fig. 3.8 which was recorded by dissolving the IF_5 in a mobile solvent (CH_2Cl_2) clearly demonstrates the difference. In the case of the neat liquid spectrum the individual components of the multiplet peaks in the spectrum are only just discernible.

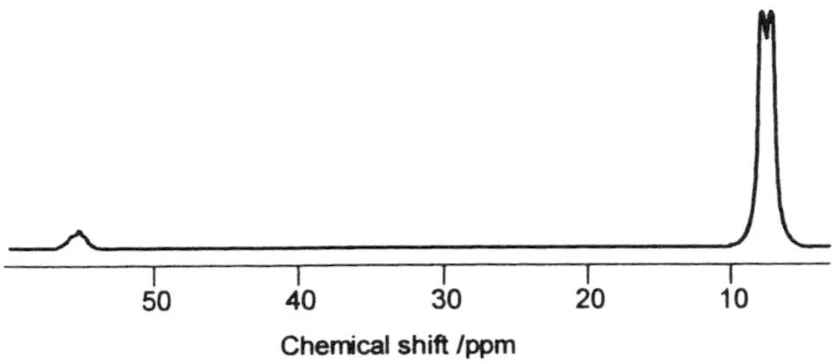

Fig 3.37 The ^{19}F NMR spectrum of neat IF_5.

The second relaxation process is by spin–spin interactions in which energy is transferred to other spin systems within the molecule. Some nuclei naturally relax more slowly than others. This is especially true for ^{13}C nuclei when there are no other spin-active nuclei, e.g. protons, attached to aid in spin–spin relaxation. For this reason tertiary carbon atoms relax more slowly than secondary than primary. This is one of the reasons that integration of carbon-13 NMR spectra is not possible.

On the other hand, the presence of paramagnetic species or quadrupolar nuclei can result in very effective relaxation of the excited spin-state and hence a very short lifetime for the excited state. This will result in the observation of broad peaks in spectra due to the Heisenberg uncertainty principal. The peaks may be so broad that they are not discernible at all, but on other occasions paramagnetic complexes such as [Cr(acac)$_3$] are deliberately added to samples so that nuclei relax more quickly.

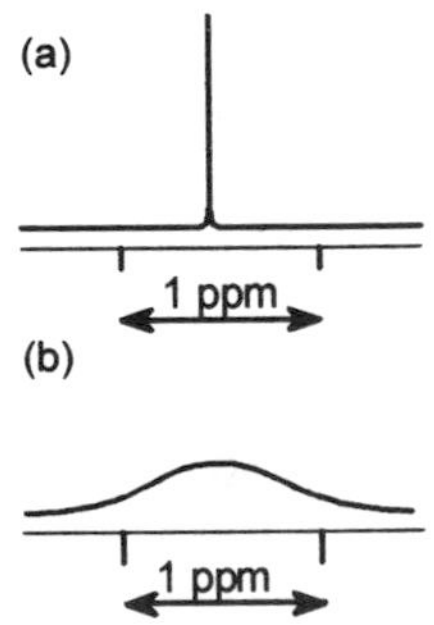

Fig. 3.38 Calculated ^1H NMR spectra for (a) NH_4^+ and (b) CH_3CN.

The presence of a quadrupolar nucleus does not always result in a broadening of peaks in the spectrum. If it is in a highly symmetric environment (octahedral or tetrahedral local symmetry), then the relaxation process is less efficient and sharp peaks are once again observed. For example, the half-height width of the proton NMR peak due to NMe_4^+ is about 0.1 Hz while that for CH_3CN is nearer 80 Hz wide. Although both of these compounds contain ^{14}N which is a quadrupolar nucleus ($I = 3/2$), the effect of quadrupolar broadening is not obvious in the first example because that nucleus is in a tetrahedral environment.

Other NMR experiments

There are a very large number of special NMR experiments available, usually based on a number of different excitation pulses. Although the theory behind these additional processes is beyond the scope of this book, two warrant a brief mention since they can greatly simplify the task of assigning peaks in complicated spectra. The first of these is called *distortionless enhancement by polarisation transfer* (abbreviated to DEPT). Such spectra are widely used in ^{13}C NMR studies to distinguish between ^{13}C nuclei in CH or CH_3 groups because these give rise to negative peaks while positive peaks result from ^{13}C nuclei in CH_2 groups.

For example, the ^{13}C NMR spectrum of $Cl(CH_2)_3Si(OCH_3)_3$ contains signals due to four different environments, one for each of the three different CH_2 carbon nuclei and one environment arising from the three equivalent OCH_3 groups. It is not immediately possible to assign the peaks in the ^{13}C spectrum, Fig. 3.39(a), unambiguously since the carbons in both the $ClCH_2$– and –OCH_3 environments are expected to resonate with similar chemical shift values. Assignment is also made more difficult by the fact that ^{13}C NMR spectra cannot be integrated with any degree of confidence. However, the DEPT spectrum in Fig. 3.39(b) clearly identifies the peak at *ca.* 50 ppm as that arising from the CH_3 environment whilst all the remaining peaks are positive and must therefore be due to CH_2 environments.

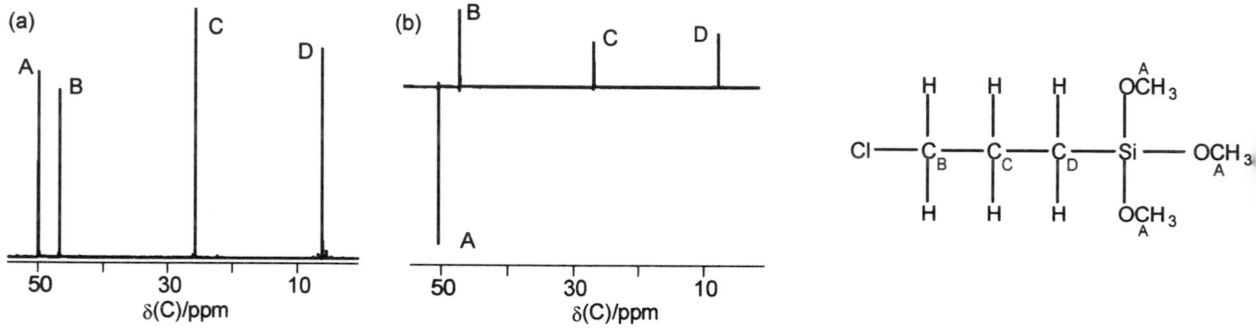

Fig. 3.39 (a) The $^{13}C\{^1H\}$ NMR and (b) DEPT spectra of $Cl(CH_2)_3Si(OCH_3)_3$.

Question 3.11 Is there any other information that allows us to be confident in the assignment of signal D in Fig 3.39 to the CH_2 bound to the silicon atom?

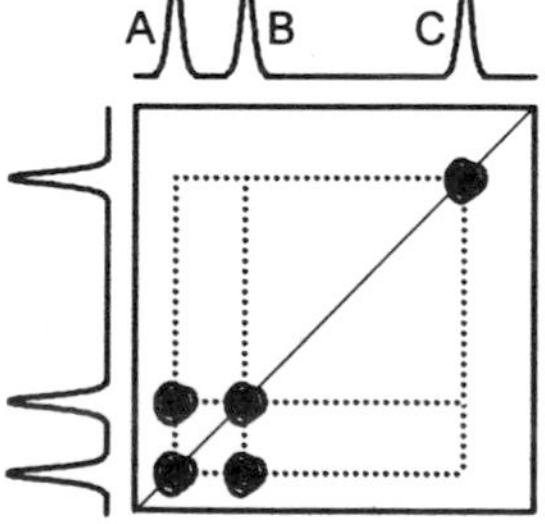

Fig. 3.40 Part of a COSY experiment.

So far all the spectra displayed are one-dimensional, that is peak-intensity is plotted against one other dimension, which is usually the chemical shift. A COSY (or *correlation spectroscopy*) spectrum has two axes each containing a spectrum. The central contour map contains peaks along the diagonal corresponding to each peak in the one-dimensional spectrum. Off-diagonal peaks indicate sets of nuclei that are coupling to each other. Such experiments can either be recorded for two similar nuclei and thus show homonuclear coupling or for two heteronuclei and identify heteronuclear coupling. By this method it is possible to identify which signals correspond to nuclei coupling with each other.

Figure 3.40 displays a typical COSY spectrum (in this case from a ^{19}F–^{19}F COSY experiment). The plot is symmetric with the ^{19}F NMR spectrum running along each axis and the central diagonal. The off-diagonal peaks link resonances which link peaks on the diagonal indicating coupling nuclei. In this case the nuclei labelled A and B are coupling but neither A nor B couple with the resonance marked C.

Magic-angle NMR

By far the most common physical state for NMR investigation is the liquid or solution phase; however, solids can also be studied using a variation of the technique. Because in solids the atoms and molecules are locked into a certain configuration, the spin–lattice relaxation time is very long, and this combined with the large number of potentially different couplings for anisotropic systems results in broad peaks. However, the term giving rise to this broadening has an angular dependence and by spinning the solid sample very rapidly at an angle of 54.7° to the magnetic field (the so-called *magic angle*) this contribution is removed resulting in narrow lines.

For example, the magic-angle ^{31}P NMR spectrum of solid-phase PCl_5 consists of two equally intense sharp lines at 86 and –295 ppm. This is at variance with the solution-phase spectrum of PCl_5 recorded in CS_2 which shows just a single peak at 80 ppm. The two peaks in the solid NMR spectrum are too far apart to be a doublet, and anyway coupling to chlorine (^{35}Cl, $I = 3/2$) would give rise to more lines. The two peaks must therefore be due to two different phosphorus chemical environments which suggests, since there is only one phosphorus atom in PCl_5, the presence of two different species. In the solid phase PCl_5 forms an ion-pair $[PCl_4]^+[PCl_6]^-$ which results in two different phosphorus environments and therefore two peaks in the solid-phase NMR spectrum.

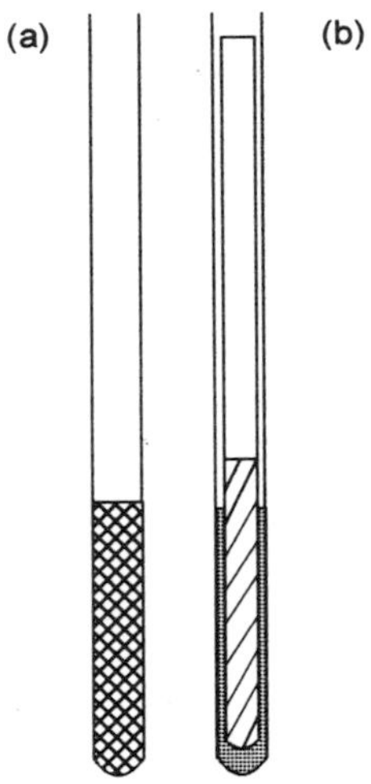

Fig. 3.41 The arrangement of NMR tubes with (a) internal and (b) external lock substances.

Experimental considerations

As was pointed out earlier, compared with many other spectroscopic techniques, NMR is fairly insensitive and this arises from the very small difference in population of the upper and lower energy states. This frequently means that larger samples are required in order to record NMR spectra than, for example, IR spectroscopy. Spectra are usually recorded of solutions in thin-walled 5-mm-diameter Pyrex tubes, but larger diameter tubes (10 mm) may be used for less sensitive nuclei so that more sample is present. Typically, a solution containing 5–10 mg of compound is required for ^{1}H and ^{19}F NMR studies, whereas 50–100 mg, or more, may be necessary for less sensitive nuclei such as ^{13}C and ^{31}P, and for very insensitive nuclei neat samples may be needed.

Deuterated solvents are most often used, partly to avoid the resonances from protons of the solvent from swamping the proton spectrum of your compound. But more importantly, the resonance of the spin-active deuterium nucleus is used as a reference to ensure that the frequencies generated throughout the experiment remain constant or 'locked'. Because the solvent is usually in excess, weak peaks may arise from small amounts of non-deuterated solvent molecules. The ^{1}H and ^{13}C positions, and multiplicities, of some of these are given in Table 3.6.

Solvent	δ(H) (ppm)	δ(C) (ppm)
$CDCl_3$	7.26 [1]	77 [3]
CD_2Cl_2	5.32 [3]	54 [5]
CD_3CN	2.08 [5]	118 [1], 1 [7]
D_2O	4.8 [1]	–
DMSO	2.49 [5]	39 [7]

Table 3.6 Resonances for the residual protonated species in some common deuterated solvents. The multiplicity of each absorption is given in brackets.

A small amount of a reference compound may also be added to act as an *internal standard* to ensure that all chemical shifts are correctly referenced. If the reference might react with your compound then an *external standard* may be used either by placing a small sealed tube of a reference material inside the sample or by adding the material into the thin gap between an inner tube and the outer glass tube. This arrangement is shown in Fig. 3.41.

3.2 Nuclear quadrupole resonance (NQR)

A nucleus with a magnetic quantum number greater than ½ will possess a quadrupolar moment in addition to a magnetic dipole moment and some examples of such nuclei have already been mentioned in the earlier section of this chapter covering NMR spectroscopy. Despite there being a relatively large number of nuclei possessing a quadrupole, the study of these systems using NQR as opposed to NMR spectroscopy is very much more limited, with probably ^{35}Cl nuclei having received the most attention.

A dipole exists when two different charges, etc., are separated, a quadrupole occurs when separation of four different charges, etc., occurs.

In NMR experiments an external field is applied to cause a splitting of the normally degenerate nuclear spin states, but this is not necessary in NQR experiments because when there is an asymmetric charge distribution within the molecule, a molecular *electric field gradient* (efg) is generated. This will arise if, for example, the molecule is composed of atoms of varying electronegativity, as shown in Fig. 3.42. The efg will interact with the nuclear quadrupolar moment resulting in a splitting of the otherwise degenerate quadrupolar states, and this is the basis of NQR spectroscopy. An efg will not, in general, be present in solutions or liquids due to molecular tumbling resulting in an averaging of the efg to zero. NQR is therefore only applicable to solids in which the local environment is not highly symmetric. The energies of the levels involved in NQR spectroscopy are given by eqn 3.9

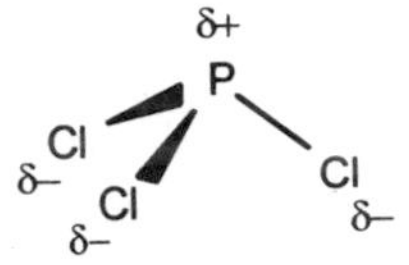

Fig 3.42 The electric field arising from the different electronegativity of the elements in the compound PCl_3.

$$E = \frac{e^2Qq[3m^2 - I(I+1)]}{4I(2I-1)} \tag{3.9}$$

where e is the charge of the electron, Q the quadrupole moment, q the field gradient, I the nuclear magnetic quantum number and m the particular nuclear spin quantum state. For the most commonly studied nuclei, such as ^{35}Cl, ^{79}Br, and ^{81}Br, which possess a spin quantum number $I = 3/2$, m will take the values $\frac{3}{2}$, $\frac{1}{2}$, $-\frac{1}{2}$, and $-\frac{3}{2}$. However, since this term involving m is squared in eqn 3.9 just two energy levels result, Fig. 3.43, and in this case the energy difference between them is

$$\Delta E = ½e^2 Q\left(\frac{\partial^2 V}{\partial z^2}\right) \tag{3.10}$$

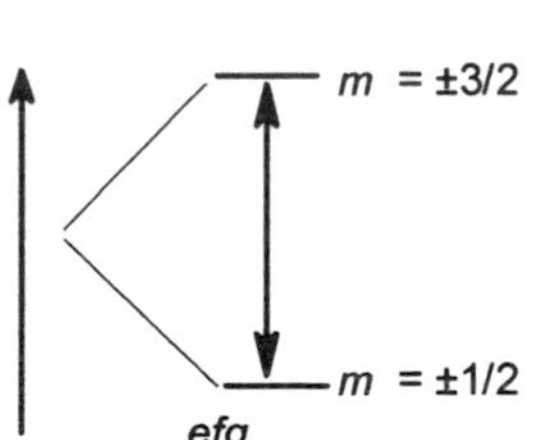

Fig. 3.43 The energy levels involved for a quadrupolar nucleus ($I = \frac{3}{2}$) in an electric field gradient.

where the differential term in the bracket is a measure of the electric field gradient in the z-direction. This energy difference usually falls in the radio-frequency part of the electromagnetic spectrum.

In NQR spectroscopy, like NMR, separate signals are observed for each set of chemically distinct nuclei under study, but in this case arising from their differing electric field gradients. However, unlike NMR spectroscopy there is no coupling between these different environments and so all peaks are observed as single absorptions. This results in the most straightforward application of NQR—one peak will be produced for each chemical environment of the nucleus under study. But even here care needs to be exercised! For example, ^{35}Cl NQR studies of solid $GeCl_4$ exhibit up to four resonances, one for each chloride ligand because low site-symmetry of the molecule in the crystal results in every chloride becoming non-equivalent. In favourable cases, however, band counting alone can allow different isomers to be distinguished. The compound $PFCl_4$ can exist in two different isomeric forms as shown in Fig. 3.44. For the first isomer with the fluoride ligand in an axial position there are two different chloride environments (axial and equatorial) in the ratio 1:3. For the second isomer two signals would also be expected, reflecting the two different chloride environments, but now their relative intensities would be in the ratio 2:2 reflecting the number of equatorial and axial chloride ligands. We should therefore be able to distinguish between these two possibilities using ^{35}Cl NQR spectroscopy.

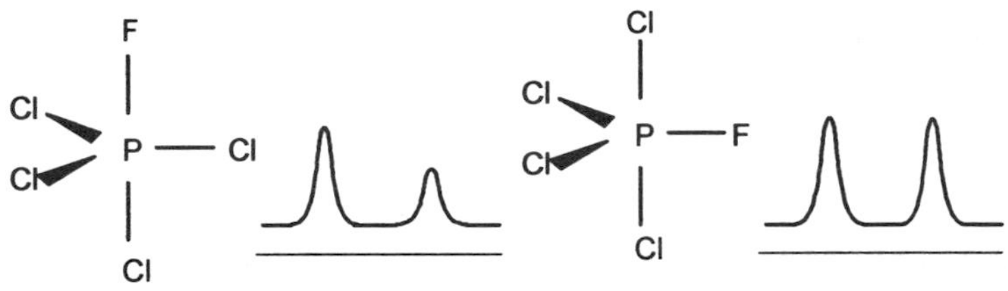

Fig. 3.44 The structures of two isomers of $PFCl_4$ and their predicted NQR spectra.

The observed spectrum displays two resonances in the ratio 1:3 suggesting that the first structure is adopted in the solid state.

A further use of NQR spectroscopy is that the resonance frequency of a particular quadrupolar nucleus can be correlated to the ionic nature of the bonding in the molecule being studied. The data in Table 3.7 show how the value of e^2Qq vary for some halogen-chlorides and alkali-metal chlorides. As the electronegativity difference increases between the constituent elements and the bonding becomes more ionic, the values become smaller. In the free chloride ion all three p-orbitals will be equally filled with two electrons and so the electron distribution is as symmetric as possible resulting in small e^2Qq values.

Question 3.12 Can NQR spectroscopy be used to distinguish between *cis*- and *trans*-$PtCl_2(NH_3)(H_2O)$?

	e^2Qq		e^2Qq
FCl	–146	KCl	0.04
BrCl	–103	RbCl	0.77
ICl	–83	CsCl	3.0

Table 3.7 NQR parameters obtained for some chlorine-containing species.

3.3 Electron spin resonance (ESR)

Just as a nucleus may possess a magnetic moment, so an electron, by virtue of the fact that it is a moving charge, will generate a magnetic moment. There are two possible values for this magnetic moment, $m_s = \pm\frac{1}{2}$. The magnetic moment is much stronger than that generated inside the nucleus and the electron is much smaller than the nucleus so the energy gap between the ground and excited state is much greater than it is for the nucleus. This larger energy gap changes the Boltzmann distribution between the ground and excited states which makes ESR spectroscopy a very much more sensitive technique than NMR spectroscopy. Since no two electrons may possess the same four quantum numbers, ESR transitions can only occur in systems with one or more unpaired electrons. The most commonly studied compounds are radical ions and transition metal complexes. Because such systems are usually paramagnetic, an alternative name for this technique is electron paramagnetic resonance or EPR.

The basis of the experiment is that when a material containing unpaired electrons is placed in a magnetic field (B_0) then splitting of the electron-spin energy levels arises (as shown in Fig. 3.45) given by

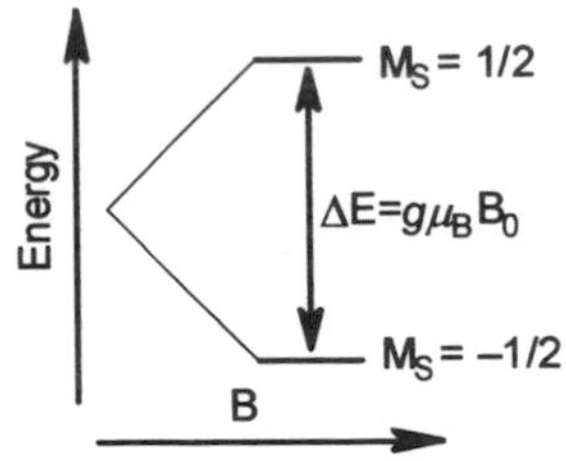

Fig. 3.45 The energy levels of an unpaired electron in a magnetic field.

$$\Delta E = g\mu_B B_0 \tag{3.11}$$

where μ_B is the Bohr magneton (9.274×10^{-24} J T^{-1}) and g is a factor of proportionality. This is equal to 2.00232 for a free electron, i.e. one that is not particularly associated with any orbital within a molecule. Most organic radicals have g-values between 1.99 and 2.01 while for transition metal species the range is usually much wider.

As we saw earlier, for NMR spectroscopy it is possible to record spectra at fixed magnetic field or fixed frequency. The same is true for ESR spectroscopy, with most ESR spectrometers working at a fixed frequency and varying the magnetic field. However, as a consequence of the way the ESR spectrometer works it is usual for spectra to be displayed in the form of the first derivative of the more usual absorption peak plotted against magnetic field. The derivative curve corresponds to a measure of the gradient of the original peak. As the absorption increases the derivative increases, but as the maximum absorption is approached the rate of change (or gradient) of the absorbance peak starts to fall until at the absorption maximum the rate of change is zero and the derivative passes through zero. These two waveforms are shown in Fig. 3.46.

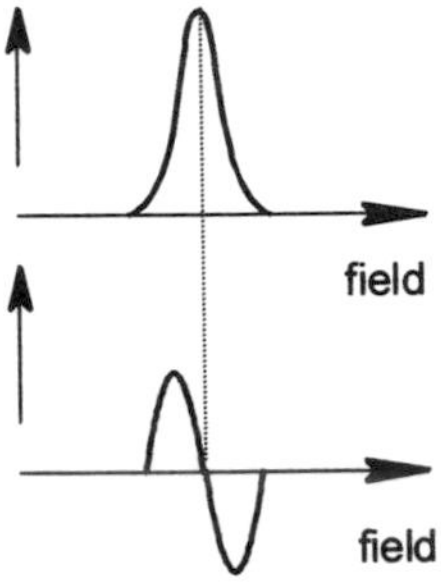

Fig. 3.46 An absorption peak and its first derivative.

Electron–electron coupling

In ESR spectroscopy coupling can arise from two different sources. Firstly, if a molecule contains more than one unpaired electron (such as transition metal complexes) then a number of possible spin-states arise. For example, if there are two unpaired electrons then the spin of the electrons may be aligned parallel or anti-parallel in three different ways (↑↑, ↑↓, or ↓↓) corresponding to $M_s = 1$, 0, and –1. Where M_s is the sum of individual m_s states. These three components would be split in a magnetic field and because of the different energies of the different states a number of transitions will be observed in the ESR spectrum. This makes interpretation of the spectrum without the aid of computer simulation programs almost impossible and certainly beyond the scope of this book and so we shall confine our discussion to systems which contain a single unpaired electron.

Question 3.13 The ESR spectrum of $ClO_2^{\bullet}$ recorded at 9136 MHz shows four lines centred at 0.325 T. What is the *g*-value?

Hyperfine coupling

Just as in NMR it is possible for one nuclear spin to interact with another nuclear spin, so it is also possible for the spin of the electron to interact with the spin of surrounding nuclei. This coupling, referred to as *hyperfine coupling*, operates in a similar way to the coupling in NMR. The ESR spectrum of a single unpaired electron would normally result in just one signal corresponding to the transition given in eqn. 3.11. If the electron is surrounded by n spin-active nuclei each with a spin quantum number of I, then a $(2nI + 1)$ line pattern will be observed in a similar way to NMR spectroscopy. For spin-½ nuclei the relative intensities of the lines will follow the binomial distribution just as the coupling patterns do in NMR spectroscopy. The energy separations are now given by

$$\Delta E = g\mu_B B_0 + m_i A_i \qquad (3.12)$$

where A_i is the *hyperfine splitting constant* and m_i the electronic spin state of nucleus i.

Figure 3.47(a) shows the ESR spectrum for the methyl radical which is a four-line pattern in the intensity ratio of 1:3:3:1. These lines arise from the interaction of the unpaired electron with the three spin-½ protons in a similar way to that observed in NMR spectroscopy. The appropriate energy diagram is shown in Fig. 3.47(b). The hyperfine splitting can be determined (in this case as 23 mT) by measuring the separation between two peaks.

Coupling to quadrupolar nuclei will also occur. For example, the ESR spectrum of the complex [Cu(acac)$_2$] shown in Fig. 3.48 exhibits four peaks in a 1:1:1:1 ratio arising from coupling of the unpaired d-electron with the ^{63}Cu nucleus which has $I = 3/2$.

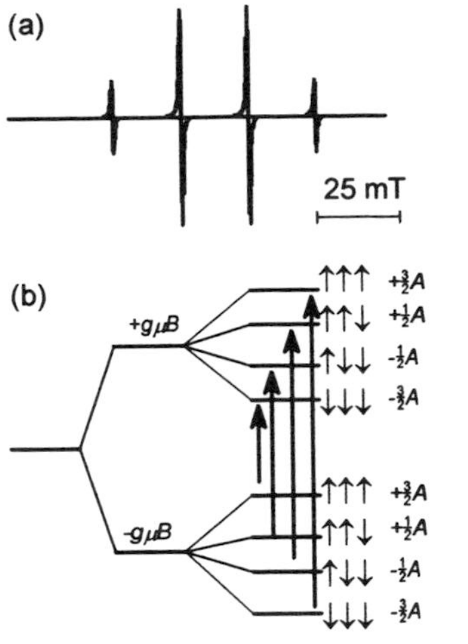

Fig. 3.47 (a) The ESR spectrum of the $CH_3^{\cdot}$ radical and (b) the energy diagram for this system.

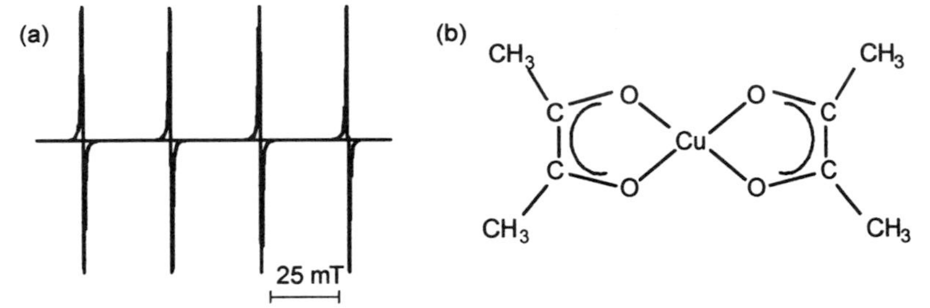

Fig. 3.48 (a) The ESR spectrum and (b) the structure of the complex [Cu(acac)$_2$].

The previous spectra are typical of those obtained from solution-phase studies. Under these circumstances the rapid tumbling of the molecules ensures that any distinction between the various cartesian directions of the molecules is averaged to zero. However, ESR studies of solids or frozen solutions are frequently recorded and under these conditions it is possible to observe two or three different sets of peaks depending on the molecular symmetry. If the compound under study possesses an axis of symmetry then the electric field in this direction is different from the remaining two and the ESR spectrum reflects this in that two separate peaks are observed corresponding to the different *g*-factors for the axis direction (labelled $g_\parallel$) and the other perpendicular directions (labelled $g_\perp$). Alternatively, all three axis directions may be unique resulting in the presence of three peaks corresponding to different *g*-factors, labelled as g_{xx}, g_{yy}, and g_{zz}. In this case three separate peaks are observed in the ESR spectrum as shown in Fig. 3.49, and each may show hyperfine coupling.

Question 3.14 Would you expect to observe ESR spectra of the following species and if so what will be their appearance: (i) I_2, (ii) I_2^-, (iii) NO, and (iv) VO^{2+}?

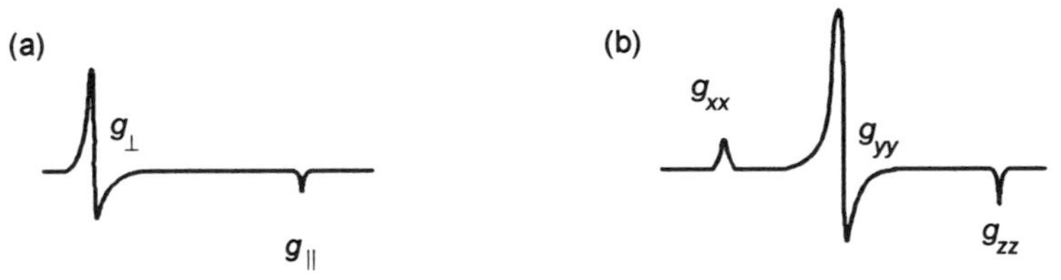

Fig. 3.49 ESR spectra of solids in which (a) $g_\perp > g_\parallel$ and (b) $g_{xx} > g_{yy} > g_{zz}$.

3.4 Potential problems

The following section identifies some of the problems encountered in NMR spectroscopy and suggests some solutions or remedial action.

Problem	Possible cause	Suggestion
Too many peaks in the spectrum	Mixture of compounds Chemical or magnetic non-equivalence	Effect separation or use 2D techniques, if appropriate.
Too few peaks in the spectrum	Inter- or intramolecular exchange processes	Record spectrum at lower temperature or in a different solvent
Too few peaks in a multiplet	(a) Weakest lines of a binomial distribution not visible (b) Accidental overlap of peaks	(a) Accumulate more scans to increase signal-to-noise level (b) Record again at a different magnetic field
Peaks very broad	(a) Exchange processes occurring (b) Poor sample preparation (c) Quadrupolar nucleus present (d) Paramagnetic compound	(a) Record spectrum at lower temperature or in a different solvent (b) Use a mobile solvent and ensure no solids are present in the tube
Strange non-binomial patterns observed	(a) Second-order spectra (b) Accidental overlap of patterns	Record on higher field machine or simulate pattern by calculation

3.5 Further questions

3.15 The ESR spectrum of the NH_2 radical is shown in Fig. 3.50. Explain the observed pattern and determine the *g*-value and the hyperfine splitting constants for this species.

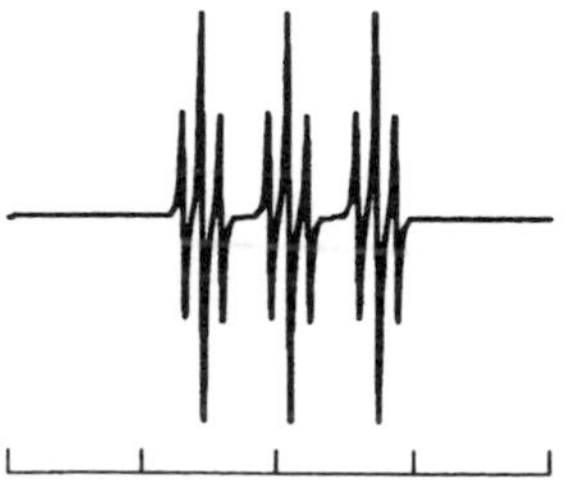

Fig. 3.50 The ESR spectrum of the NH_2 radical.

3.16 The ^{31}P NMR spectrum of $[Rh(PEt_3)_3Cl_2Br]$ exhibits a doublet of doublets centred at –4.3 ppm and a doublet of triplets at –20 ppm. Explain whether these patterns allow the stereochemistry of the complex to be unambiguously determined, if not what experiment might?

3.17 Identify the white product from the reaction of $[Pt(PPh_3)_2(CO_3)]$ with $NaBH_4$ and C_2H_4 which contains 60.8% C and 4.65% H. Room temperature multinuclear NMR experiments reveal a number of peaks. Of these the following show platinum satellite structure: ^1H, 2.2 ppm (singlet); $^{13}C\{^1H\}$, 39.6 ppm (singlet). The ^{195}Pt NMR spectrum displays a triplet.

Further reading

P.J. Hore, Nuclear magnetic resonance (Oxford chemistry primers series), Oxford University Press (1995).

R.V. Parish, NMR, NQR, EPR and Mossbauer spectroscopy in inorganic chemistry, Ellis Horwood (1990).

A.E Derome, Modern NMR techniques for chemistry research, Pergamon Press (1987).

R.K. Harris and B.E. Mann, NMR and the periodic table, Academic Press (1978).

4 UV–visible spectroscopy

4.1 Introduction

Within inorganic chemistry the field of study most often associated with UV–visible spectroscopy is that of the coloured transition metal complexes. The fact that complexes of certain transition metals in particular oxidation states are of a similar colour is well known. For example, most copper(II) complexes are blue and many iron(II) and nickel(II) complexes are green which suggests that the colour arises from the metal. However, the observed colour can change quite dramatically on altering the oxidation state of the metal; for example, green coloured iron(II) complexes become orange/brown in analogous iron(III) complexes. This is because the colour is dependent on the metal involved and the number of d-electrons it possesses, which, in turn, is related to its oxidation state. However, some main group compounds and transition metal complexes which do not contain d-electrons are also coloured, in fact very intensely—this colour also arises from electronic transitions, but involving other valence electrons.

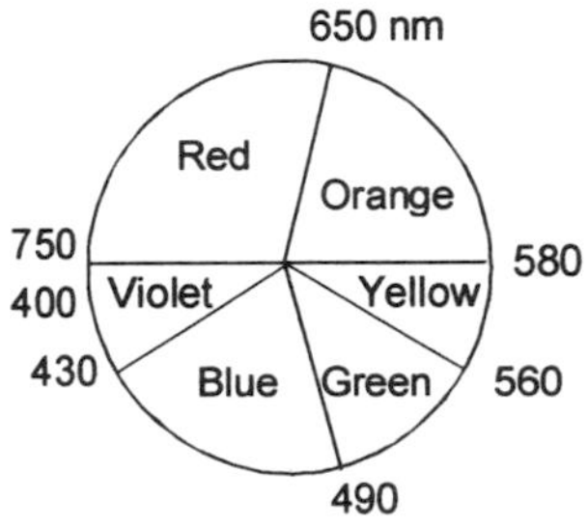

Fig. 4.1 A colour wheel—the absorption of certain energies of white light results in the complementary colour being observed.

The energies associated with transitions between different arrangements of valence electrons falls within the ultraviolet (UV) and visible region of the electromagnetic spectrum. Just as the most popular form of vibrational spectroscopy is usually referred to by the region it occurs in—the infrared—so the most popular form of electronic spectroscopy is usually known as UV–visible spectroscopy and is in essence the study of the transitions involved in the rearrangements of valence electrons.

4.2 Experimental methods

Electronic spectroscopy is an absorption technique and the colour observed for a complex is determined by which frequencies of visible light are *not* absorbed. If white light falls on a particular complex which absorbs energies at the red end of the visible spectrum (*ca.* 750 nm, Fig. 4.1) then the complex appears to be blue, that is we see the *complementary colour*. In a similar way if a sample absorbs radiation corresponding to all visible colours then it appears to be black.

Material	Cut-off (nm)
Glass	300
Quartz	170
Water	190
Acetone	330
Ethanol	205
Chloroform	250

Table 4.1 The UV cut-off for some solvents and cell materials.

Most electronic spectroscopy is carried out by making up a sample of known concentration in a suitable solvent (Table 4.1) bearing in mind that solvent coordination might result in the formation of a completely different complex. A portion of the solution is placed in a cell of known path length, usually 1 cm, with neat solvent in a second similar holder as a reference. The spectrum presented therefore consists of peaks due to absorption by the sample and not the solvent. Frequently, UV–visible spectra will contain bands which vary in intensity by factors of many tens or hundreds so it is often necessary to record spectra at a number of different concentrations so that none of the weaker bands are missed.

There are two important parameters to be obtained from electronic spectra; firstly the positions of the absorption peaks and secondly the intensity of the band. From the intensity and the concentration of the sample, the *extinction coefficient*, ε, for each peak can be calculated according to the Beer–Lambert Law

$$A = \log_{10}(I_0/I) = \varepsilon\, cl \tag{4.1}$$

where A is the absorbance, c is the molar concentration and l the path length in centimetres.

Question 4.1 A 2×10^{-3} mol dm^{-3} concentration solution of a compound in a 1 cm path length cell results in an absorption of 1.5. What is the material's extinction coefficient?

Electronic transitions are extremely rapid, typically occurring in 10^{-15} seconds and according to the *Franck–Condon principle* the atoms of the molecule do not have time to change position during the transition. Since molecules are constantly vibrating it means that when an electronic transition is induced, molecules will be in many different vibrational states and therefore a range of energies is absorbed as indicated by the diagram in Fig. 4.2(a). In some cases it is possible to distinguish a number of resolved peaks on a single electronic band, such as that shown in Fig. 4.2(b) for the permanganate ion. By measuring these peak positions it is possible to obtain an estimate of the vibrational frequency of the excited state responsible for the fine structure. It is usual first to convert from nm to cm^{-1}.

In UV–visible spectroscopy the units nm and cm^{-1} are in common use; you can convert from one to the other using $\nu(\text{cm}^{-1}) = 10^7/\lambda(\text{nm})$.

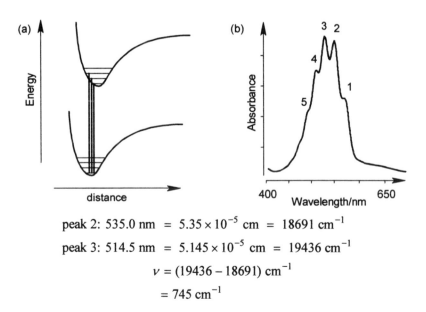

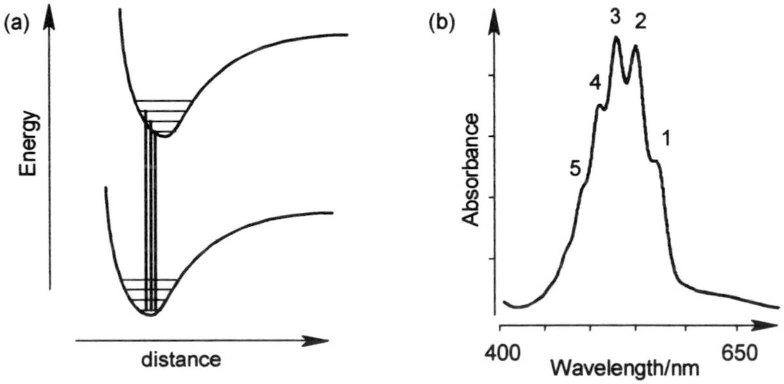

Fig. 4.2 (a) The source of vibrational fine structure on electronic bands; (b) for $[MnO_4]^-$ these are well resolved.

peak 2: 535.0 nm = 5.35×10^{-5} cm = 18691 cm^{-1}

peak 3: 514.5 nm = 5.145×10^{-5} cm = 19436 cm^{-1}

$$\nu = (19436 - 18691)\ \text{cm}^{-1}$$
$$= 745\ \text{cm}^{-1}$$

However, in the majority of cases the vibrational fine structure is not sufficiently well resolved and so broad peaks (of the order 1000 cm^{-1} wide) are observed in the UV–visible spectrum.

4.3 Metal–metal transitions

The colours most often associated with transition metal complexes arise from transitions between different energy levels corresponding to a redistribution of electrons in the partially filled d-orbitals. These are referred to as metal–metal or d–d transitions. However, before discussing such transitions in detail we need to consider the organisation of the electrons in the various d-orbitals. Representations of these orbitals are shown below in Fig. 4.3.

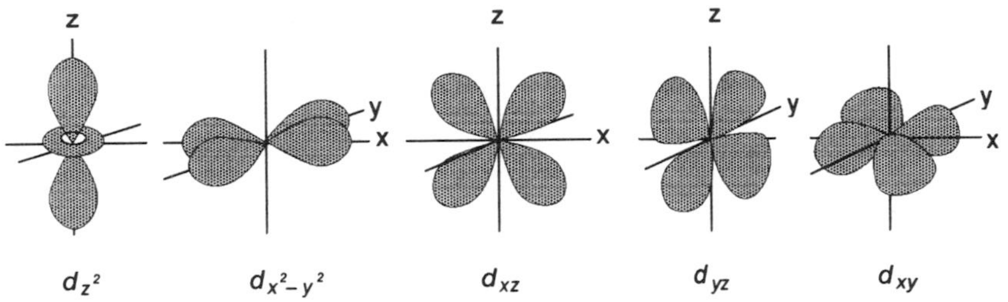

Fig. 4.3 Representations of the d-orbitals.

Crystal field splitting

In the absence of any ligands around the metal, the energies of all five d-orbitals of a transition metal are equal, that is they are degenerate. The presence of ligands will, based on a purely electrostatic model, result in an increase in the energy levels of all these orbitals. Anything but a completely spherical distribution of electron density evenly spaced around the metal will result in removal of the degeneracy, because some ligands and orbitals will interact more strongly than others. The exact form of the interaction and hence energies of the d-orbitals depend on the type, number, and spatial arrangement of the ligands.

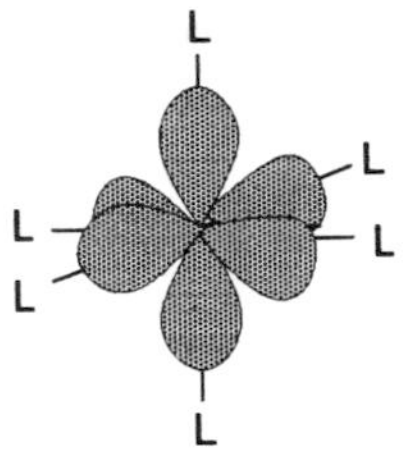

Fig 4.4 The six ligands in an octahedral arrangement interact most strongly with the d_{z^2} and $d_{x^2-y^2}$ orbitals.

The most common geometry encountered in transition metal chemistry is based on octahedral symmetry with the six ligands lying along the *x*, *y*, and *z* axes. In this arrangement the greatest interaction will occur between the ligands and the metal $d_{x^2-y^2}$ and d_{z^2} orbitals since these orbitals have their maximum electron density along the *x*, *y*, and *z* axes, as shown in Fig. 4.4. The d_{xy}, d_{xz}, and d_{yz} orbitals have their electron density maxima lying between the axes and so interaction between these orbitals and the ligands will be less.

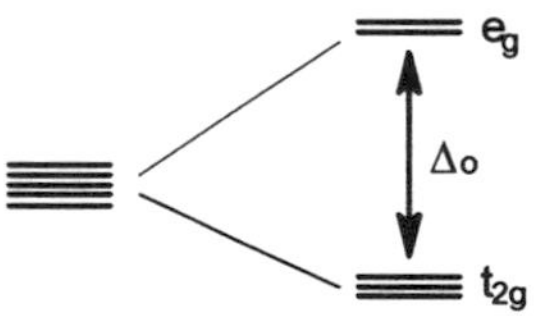

Fig 4.5 The energy-level splitting diagram for octahedral geometry.

The presence of six octahedrally coordinated ligands therefore results in the removal of the fivefold degeneracy of the metal d-orbitals resulting in two sets of orbitals, one triply degenerate and the other doubly degenerate, as shown in Fig 4.5. These are referred to by their symmetry labels—the three lower energy orbitals transform as t_{2g} and the upper pair as e_g under the octahedral point group (O_h). The 'g' subscript coming from the German word *gerade*, meaning even, refers to the fact that each orbital, or set of orbitals, are symmetric about the centre of symmetry in the molecule. Orbitals which are not centrosymmetric are labelled by the letter 'u' for *ungerade*, or uneven.

The sets of orbitals are separated by an energy gap labelled as Δ_o (the subscript 'o' refers to the octahedral geometry), although in some texts this

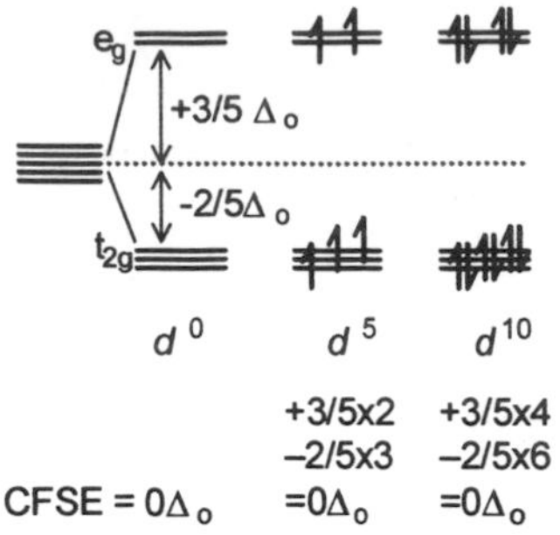

Fig. 4.6 The crystal field splitting energy arising from zero, singly, and doubly occupied d-orbitals of an octahedral complex.

energy gap is referred to as 10Dq. The splitting of the energy levels is weighted by the number of orbitals involved, so that when the orbitals are zero, singly, or doubly occupied there is no net energy change, as shown in Fig. 4.6. The t_{2g} orbitals are more stable by 2/5 Δ_o and the e_g orbitals less stable, i.e. at higher energy, by 3/5 Δ_o. This splitting of the orbital energy levels is referred to as *crystal field splitting* when an electrostatic model (as described above) is used. By multiplying the number of electrons in each orbital and its relative energy, expressed as a function of Δ_o, the *crystal field stabilisation energy* or CFSE is obtained. A more rigorous treatment of this problem would consider other properties of the ligand, such as its overall size and whether it can participate in π-bonding, and when such factors are taken into account we refer instead to *ligand field splitting.*

The actual magnitude of the splitting and hence Δ_o will depend on a number of factors including the size and charge of the metal ion and ligands involved. For these reasons Δ_o for a 4d metal is about 50% larger than for a 3d metal and for a 5d metal a further 25% larger than a 4d metal. The effect that varying the ligand has on Δ_o is dealt with in more detail later.

The energy gap Δ_o is of the same magnitude as the spin-pairing energy—the energy difference between two electrons paired with anti-parallel spins in the same orbital compared with them being in two different orbitals with parallel spin. This results in the possibility of two different electronic configurations arising for many d^n electronic configurations. Figure 4.7 shows two possible arrangements for a d^4 electronic configuration; the first of these corresponding to $(t_{2g})^4$ is called the *low-spin* configuration since there are fewer unpaired electrons than in the second, alternative, *high-spin* configuration of $(t_{2g})^3(e_g)^1$. Bearing in mind the relative magnitude of the t_{2g}–e_g energy gap outlined above for the 3d, 4d, and 5d metals we should expect 3d transition metal complexes to be the most likely to adopt high-spin configurations.

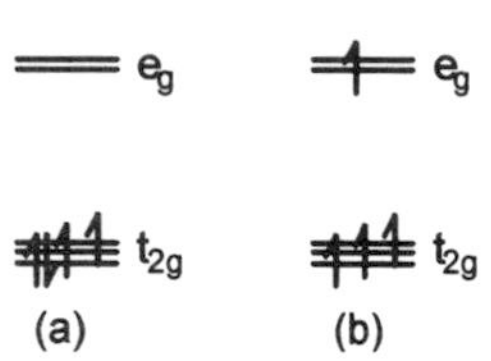

Fig. 4.7 (a) Low- and (b) high-spin configurations for a d^4 transition metal ion.

Crystal field splitting in other common shapes

Although the octahedral arrangement of six ligands around a metal centre is the most prevalent for transition metal complexes, tetrahedral and square-planar geometries are also fairly common. In the case of a tetrahedral complex the arrangement of ligands and orbitals is now such that there will be a lesser electrostatic interaction between the ligands and the d_{z^2} and $d_{x^2-y^2}$ orbitals than with the d_{xy}, d_{xz}, and d_{yz} orbitals, Fig. 4.8. This results in a reversal of the splitting pattern observed for complexes with octahedral symmetry and is shown in Fig. 4.9(a).

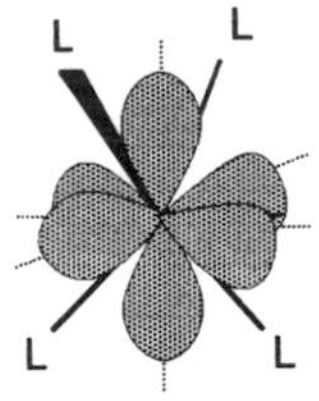

Fig. 4.8 Tetrahedrally coordinated ligands interact least strongly with the d_{z^2} and $d_{x^2-y^2}$ orbitals.

The degree of interaction between the ligands and orbitals is less in the tetrahedral case, one intuitive reason for this being that there are four rather than six ligands causing the splitting. The symmetry labels for these sets of orbitals are e and t_2 and reflect the fact that the tetrahedral point group does not have a centre of symmetry, thus removing the distinction between gerade and ungerade. The energy gap between the two sets of orbitals is labelled as Δ_t and it is just a little under half the size of the octahedral gap Δ_o as given below

$$\Delta_t = \frac{4}{9}\Delta_o \quad (4.2)$$

The second common four-coordinate geometry is the square-planar shape adopted, mainly, by a number of the heavy, late-transition metal complexes. The energy level diagram for this system is obtained by considering an octahedral complex and then removing two *trans*-ligands, which for the sake of argument we will consider to be the two on the *z*-axis. As these ligands are pulled away from the metal centre, the interaction with the metal d-orbitals possessing a *z*-component will decrease resulting in a lowering of the energy of the d_{z^2} and, to a lesser extent, of the d_{xz} and d_{yz}. To counter this the energy of the remaining orbitals will increase. As these two ligands are pulled further away along the *z*-axis, the energies of the orbitals are affected further until the ligands are completely removed resulting in a square-planar complex and the energy level diagram resembles that shown in Fig. 4.9(d).

Question 4.2 Are tetrahedral complexes more or less likely to adopt high-spin configurations than octahedral ones?

The electronic configuration resulting in the greatest crystal field stabilisation energy for such a geometry would be d^8 and as a consequence of this the majority of d^8 metal complexes adopt square-planar geometry.

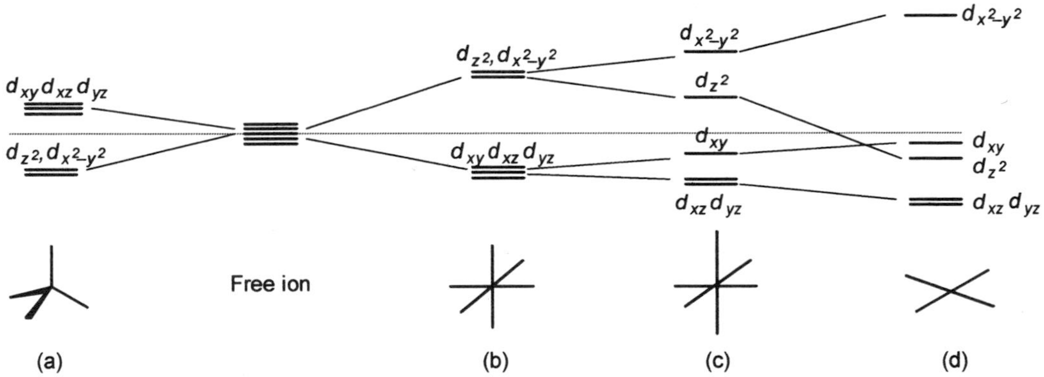

Fig 4.9. The energy level diagrams for (a) tetrahedral, (b) octahedral, (c) distorted octahedral, and (d) square-planar geometry.

Based on the splitting of these orbitals we should expect a single electronic transition for a d^1 transition metal complex of octahedral or tetrahedral geometry, corresponding to the excitation of the single electron from the ground to excited state. For a d^1 metal in distorted octahedral or square-planar geometries a larger number of transitions are possible.

Figure 4.10(a) shows the visible spectrum of the complex $[Ti(H_2O)_6]^{3+}$ which consists of a single transition at *ca.* 500 nm. This wavelength corresponds to the green part of the visible spectrum and therefore the complex appears purple in colour.

Question 4.3 How many d–d transitions would you expect for a d^1 square-planar complex?

By absorbing a photon of energy corresponding exactly to the t_{2g}–e_g energy gap, the single electron can, in theory, be promoted from the lower set of orbitals to the upper ones, as illustrated in Fig. 4.10(b). Such a transition is referred to as a d–d transition since the electron is being excited from one d-electron level to another. The frequency at which this occurs will correspond to the energy gap Δ_o.

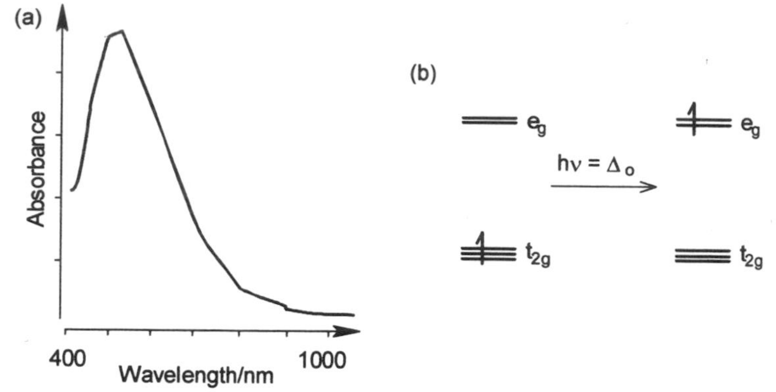

Fig. 4.10 (a) The visible spectrum for the d^1 octahedral complex $[Ti(H_2O)_6]^{3+}$ and (b) the transition responsible.

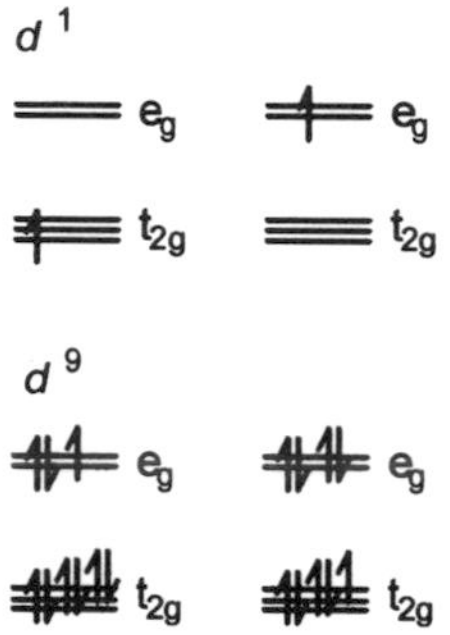

Fig. 4.11 The single 'hole' of a d^9 configuration can be thought of in a similar way to the d^1 configuration.

A single transition should also be expected for a d^9 octahedral metal ion, since d^1 and d^9 configurations are related. In the d^1 ion the single electron can occupy one t_{2g} or e_g orbital. In a similar way for the d^9 case, the 'missing' electron or 'hole' can occupy one t_{2g} or e_g orbital as shown in Fig. 4.11 and therefore there is only one possible transition, corresponding to $(t_{2g})^6(e_g)^3 \rightarrow (t_{2g})^5(e_g)^4$ For most other electron arrangements the situation is more complicated.

Spin states

The octahedral splitting of the metal d-orbitals results in a t_{2g} and an e_g set of orbitals; however, placing electrons in these levels may result in further interactions resulting in a number of different *spin states*. In the absence of any ligands around the metal, the ground state of the free ion can be written in a short-hand fashion known as a spectroscopic term symbol according to the *Russell–Saunders* coupling scheme. This scheme which uses a letter to describe the electronic state and a number to identify the spin multiplicity is appropriate for the lighter transition metal elements where coupling of the induced magnetic dipoles is small. This is not so for the heavier transition metals, where an alternative method which accounts for spin–orbit coupling, called the *jj-coupling* scheme, is more appropriate. References to texts with further details of this scheme can be found at the end of this book.

Russell–Saunders coupling scheme

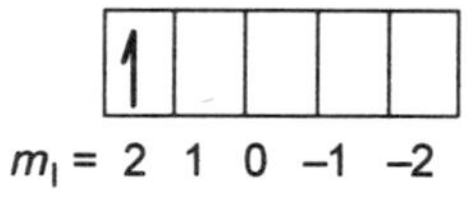

Fig. 4.12 A simple representation of the d^1 ion.

In the ground state the single electron of a d^1 metal ion, according to Hund's rules, will occupy the orbital with the highest magnetic quantum number m_l = 2, as shown in Fig. 4.12. L, the total angular momentum, given by the sum of m_l values, is therefore 2 and this state is labelled D according to the scheme given below

L	=	0	1	2	3	4	5	...
label	=	S	P	D	F	G	H	...

The spin multiplicity is given by $(2S+1)$ where S is defined as the sum of m_s values (the spin of the electron). In this case there is only one electron ($m_s=\frac{1}{2}$) so the total spin is ½ and therefore the spin multiplicity is 2. Thus the spectroscopic term symbol according to the Russell–Saunders notation for a d^1 metal ion is ^2D (which is said as doublet D). This scheme may be used to determine equivalent symbols for all the possible ground state ions, as shown in Table 4.2.

d^n configuration	Term symbol
d^1	^2D
d^2	^3F
d^3	^4F
d^4	^5D
d^5	^6S
d^6	^5D
d^7	^4F
d^8	^3F
d^9	^2D
d^{10}	^1S

Table 4.2 Spectroscopic term symbols for d^n metal ions.

However, we are concerned with the arrangement of electrons within a complex rather than the free ion. This can be obtained in two different ways depending on the relative importance of the crystal field splitting and the interelectronic interactions. In the *weak field* approach we consider the interelectronic repulsions as being the dominant effect and build their effects onto the free metal ion splitting. In the *strong field* approach the crystal field splitting is the most important factor and electronic repulsions are considered as secondary to this effect. We will concentrate on the strong field approximation.

Strong field model

Octahedral coordination of ligands to a metal results in the splitting of the degeneracy of the d-orbitals to generate t_{2g} and e_g components. In the case of a d^1 ion, since there is only one electron, there can be no interelectronic repulsions so the labels for the electronic states ^2E_g and $^2T_{2g}$ reflect the labels given to the orbitals. These states correspond to when the single electron resides in the t_{2g} or e_g set of orbitals. The superscript two refers to the spin multiplicity of the electronic states, derived in the same way as the free ion. By convention, capital letters are used to denote an electronic state while lower case letters are used to refer to orbitals. The single d–d transition shown in Fig. 4.10(a) for $[Ti(H_2O)_6]^{3+}$ is, therefore, labelled as $^2T_{2g} \rightarrow {}^2E_g$.

In some texts you will find transitions written as $^2E_g \leftarrow {}^2T_{2g}$, that is the excited state is written first. However, the alternative, more logically read version with the forward arrow is used throughout this chapter.

For a d^2 metal ion there are a larger number of potential electronic configurations to be considered. However, we will only deal with those that are the most likely since they obey Hund's rule of maximum multiplicity and the Pauli exclusion principle. Those possibilities are shown in Fig. 4.13. The first corresponds to the lowest energy ground state with the electrons residing in two of the three t_{2g} orbitals. There are three such possible permutations $[(d_{xy})^1(d_{xz})^1(d_{yz})^0$, $(d_{xy})^1(d_{xz})^0(d_{yz})^1$, and $(d_{xy})^0(d_{xz})^1(d_{yz})^1]$ all of the same energy, so this is a triply degenerate state. The total spin multiplicity for this arrangement given by $(2S+1)$ will be 3 $[2(\frac{1}{2}+\frac{1}{2})+1 = 3]$. Thus we would label this spin state as ^3T.

The next two possibilities shown in Fig. 4.13 (b) and (c) correspond to $(t_{2g})^1$ $(e_g)^1$ arrangements, which are separated on the basis that the electrons either have components in all three cartesian axes, $(d_{xy})^1(d_{z^2})^1$, $(d_{yz})^1(d_{x^2-y^2})^1$, and $(d_{xz})^1(d_{x^2-y^2})^1$ or can be confined to just two axes such as $(d_{xy})^1(d_{x^2-y^2})^1$, $(d_{xz})^1(d_{z^2})^1$, and $(d_{yz})^1(d_{z^2})^1$. It would be reasonable to assume that confining the electrons in just two dimensions rather than three is a higher energy arrangement. As can be seen from the preceding sentences, in both cases there are three possible configurations of the same energy and therefore these are triply degenerate, they both have a multiplicity of three and will both correspond to ^3T states. The difference in energy between these two states will be related to the interelectronic repulsion energetics. Both of these ^3T states differ from the ground state by the promotion of one electron from the t_{2g} set of orbitals to one of the e_g orbitals, which means they will both be at higher energy compared with the ground state. If electronic repulsions are ignored this energy difference will be equal to Δ_o.

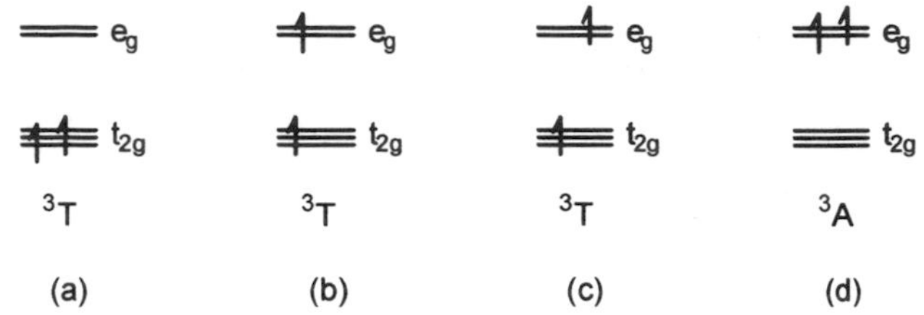

Fig. 4.13 Four possible arrangements of the electrons of an octahedral d^2 complex.

There is only one way the final representation corresponding to $(e_g)^2$ can be achieved and therefore this is a singly degenerate state and is labelled as ^3A. This arrangement cannot arise by a single electron transition from the ground state of a d^2 ion but instead will require energy equivalent to approximately $2\Delta_o$ to attain this higher energy state. Because this is a two-electron transfer process it is much less likely to occur and will therefore be weaker than the bands corresponding to single-electron processes. We should therefore expect a d^2 octahedrally-coordinated metal ion to exhibit three d–d transitions of which one will be correspond to a two-electron jump and therefore be less intense. These correspond to transitions from the ground state to any one of the other energy levels, as shown in Fig. 4.14.

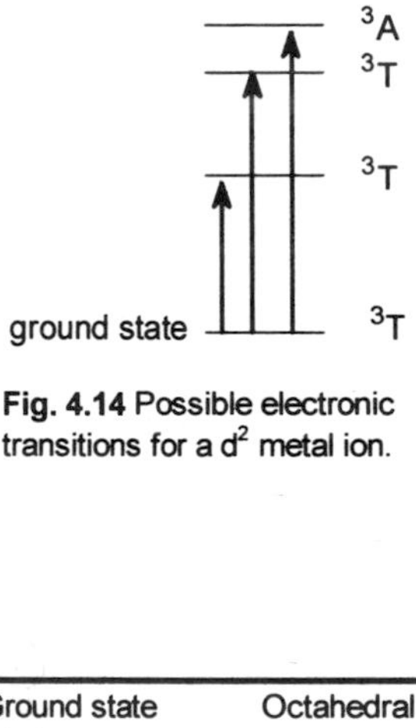

Fig. 4.14 Possible electronic transitions for a d^2 metal ion.

Just as the d^9 configuration is related to d^1 by the single electron-hole formalism, so d^2 and d^8 configurations can be considered in a similar way and so we should also expect three d–d transitions for a d^8 ion.

Weak field model

The previous discussion of the relative energy levels above assumes that the interelectronic repulsions are small compared with the crystal field splitting. The alternative approach takes the free ion as the starting point and then builds interelectronic repulsions onto that.

For a d^2 ion the ground state is labelled, according to the Russell–Saunders scheme, as ^3F. However, there are excited state arrangements that need to be considered, the derivation of which is beyond the scope of this book. For the d^2 ion the complete set of possible electronic states has the following term symbols: ^3F, ^1D, ^3P, ^1S, and ^1G. On placing the free ion in an octahedral field, splitting of these states will occur, as given in Table 4.3. Since the ground state is a triplet state the most likely transitions are those involving other triplet states, so from the above list of all possible states we just consider the ^3F and ^3P states. Using the corresponding entries given in Table 4.3 yields $^3A_2 + {}^3T_1 + {}^3T_2$ derived from the ^3F state and $^3T_1$ for the ^3P excited state.

Ground state label	Octahedral label
S	A_1
P	T_1
D	$E+T_2$
F	$A_2+T_1+T_2$
G	$A_1+E+T_1+T_2$
H	$E+2T_1+T_2$

Table 4.3 Ground state electronic configurations and their counterparts in an octahedral field.

Irrespective of which method is chosen, four triplet electronic states arise, one of which is the ground state, and so we should expect to observe three transitions. When the other d-electron configurations are considered in a similar way, the splittings shown in Fig. 4.15 are obtained from which it is expected that only a single d–d transition should be observed for d^1, d^4, d^6, and d^9 octahedral configurations, while three transitions are anticipated for d^2, d^3, d^7, and d^8 complexes. Transition metal complexes possessing a d^5 configuration are considered separately later on.

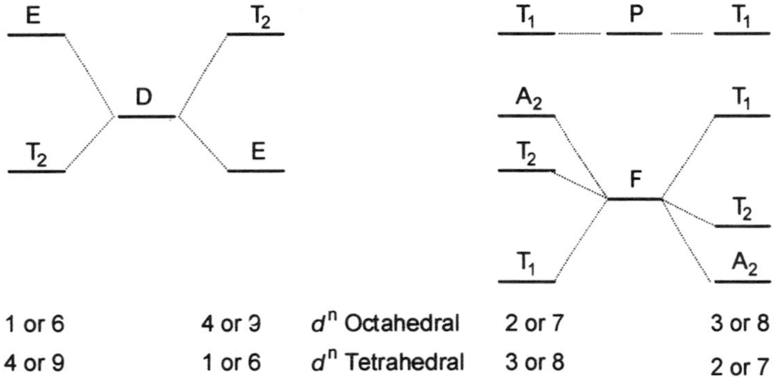

Fig. 4.15 The splitting of energy levels for high-spin octahedral and tetrahedral d^n configurations.

Although in the extreme limits of the strong and weak fields we can approximate the relative energies of these electronic states, the exact ordering of all these levels is determined by the relative magnitude of Δ_o and the electronic repulsion energies. The latter interactions are normally defined by a number of variables which are brought together in two *Racah parameters B* and *C*. For the free metal ions *B* is typically about 1000 cm^{-1} and *C* is around 4×B.

There are two common types of energy plots for electronic states: one is the Orgel diagram which shows how the electronic levels change as a function of Δ_o, such as that shown in Fig. 4.16 for the d^1 ion. These diagrams assume that the interelectronic interactions are insignificant compared with the crystal field splittings.

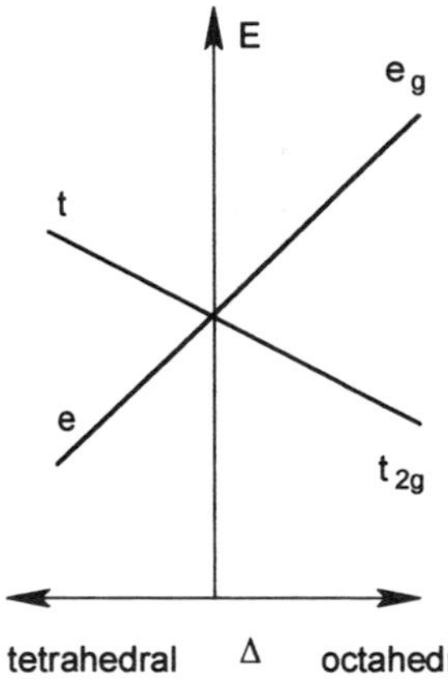

Fig. 4.16 Orgel diagram for d^1 ion.

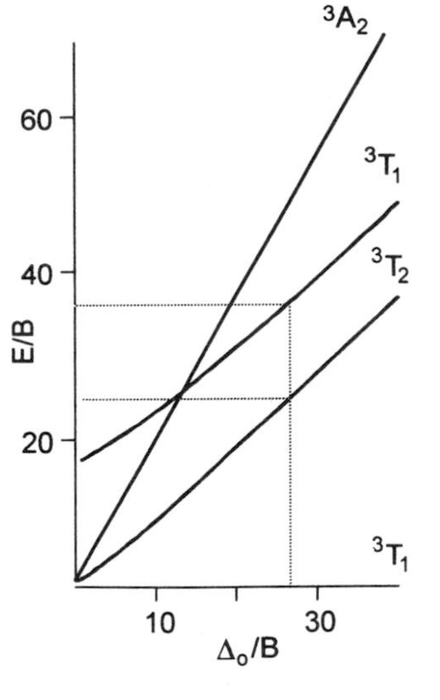

Fig. 4.17 Tanabe–Sugano diagram showing the spin-allowed transitions for the d^2 ion (plotted for *B*/*C* = 4).

The alternative Tanabe–Sugano diagram is a plot of orbital energies (relative to that of the ground state) as a function of the Racah parameter B verses Δ_o/B. Figure 4.17 shows a simplified version of such a diagram for the electronic states of the d^2 ion discussed above.

These diagrams can be used in two ways. They can either be used to obtain an estimate for the Δ_o value once d–d transitions have been identified or they may be used to predict the position of electronic transitions. For example, the three d–d transitions of the d^2 metal complex $[V(H_2O)_6]Cl_3$ are at 17400, 25600, and 38000 cm^{-1}. The ratio of the first two absorptions is 1.47; tracing along the $^3T_2$ and $^3T_1$ lines this ratio is found for the point on the horizontal scale of the Tanabe–Sugano diagram at Δ_o/B = 27. The corresponding point on the vertical axis for the lowest energy transition is at E/B=25.5. Since E, the lowest energy transition, occurs at 17400 cm^{-1}

$$E / B = \frac{17400 \text{ cm}^{-1}}{B} = 25.5$$

$$B = 680 \text{ cm}^{-1}$$

$$\text{and since } \Delta_o / B = 27$$

$$\Delta_o = 18360 \text{ cm}^{-1}$$

Question 4.4 The first two transitions of $[V(CN)_6]^{3-}$ are 22000 and 28500 cm^{-1}. Derive an estimate of Δ_o and B.

This B value is lower than the value of 1000 cm^{-1} which was previously suggested as a 'typical' value for the free ion. The ratio of B for a complex and that for the free ion is given the symbol β and is called the *nephelauxetic* parameter.

$$\beta = \frac{B(\text{complex})}{B(\text{free ion})} \tag{4.2}$$

A small β value implies that there has been some delocalisation of the electrons from the metal onto the ligand. Generally, the smaller the nephelauxetic parameter the more covalent the character of the bonding between the metal and ligand which is to be expected for softer ligands.

Question 4.5 B(free ion) for V^{3+} is 855 cm^{-1}: calculate β for the complex in Question 4.4 and comment on its value.

Ligands can be ordered on the basis (1–β) to provide the nephelauxetic series, part of which is shown below.

$$F^- < H_2O < NH_3 < Cl^- < CN^- < Br^- < I^-$$

This series lists ligands in order of increasing covalency of the bonding of the ligands to a metal. For example, amongst the halides the bonding of the fluoride ligand will be the least covalent with chloride, bromide, and iodide progressively becoming more so.

Selection rules

So far nothing has been said about the selection rules which determine which transitions are formally allowed. In electronic spectroscopy there are three. The first of these is $\Delta l = \pm 1$, that is transitions which involve a change in the quantum number l (where l = 0 corresponds to an s-orbital, l = 1 a p-orbital, l = 2 a d-orbital, etc.) by one are formally allowed, so s→p, p→d, and d→f transitions are allowed. This immediately means that d→d transitions are formally forbidden.

The second selection criterion is based on the spin-multiplicity of the states, with transitions between states of the same multiplicity being allowed but not those between different multiplicity states. Thus transitions from, say, a ^{2}T to a ^{2}E state are allowed while those in which the spin multiplicity changes such as ^{2}T to ^{1}T are not. Such a change in spin multiplicity corresponds not only to an electronic transition but also to a change in the spin of the electron involved as shown in Fig. 4.18. Transitions that are forbidden by this criteria are called *spin forbidden* and are usually very weak.

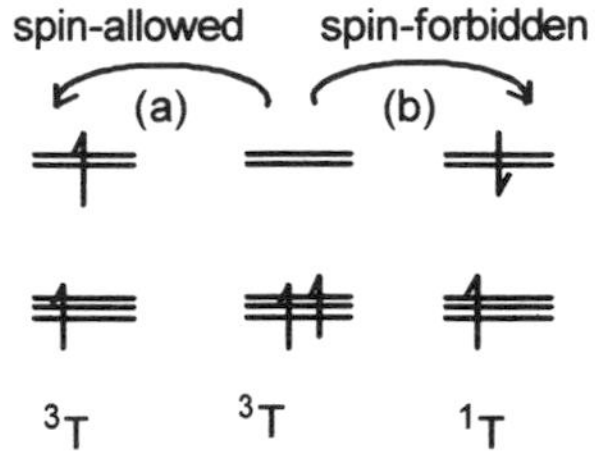

Fig. 4.18 The spin-multiplicity selection rules mean that transition (a) is favoured over that of (b).

The third selection rule, called the *Laporte* or *parity* selection rule, is based on the symmetry of the complex. For highly symmetric complexes which possess a centre of symmetry or inversion (represented by i in point group notation), such as octahedral molecules, the parity rule forbids transitions between energy levels with the same symmetry (or parity) with respect to the centre of inversion. So transitions between two gerade or two ungerade orbitals are forbidden and therefore the $^2T_{2g}\rightarrow{}^2E_g$ transition shown in Fig. 4.10 for the complex $[Ti(H_2O)_6]^{3+}$ is formally not allowed on two counts. Firstly, it is disallowed because it is a d–d transition and so $\Delta l \neq \pm 1$, and secondly, since both the ground and excited electronic states are of 'g' symmetry with respect to the inversion centre it is Laporte forbidden.

Fortunately, the Laporte rule can often be circumvented, either because the molecule is not totally symmetric due to the presence of different ligands, or because as the molecule vibrates the centre of symmetry is temporarily removed. This interaction between electronic and vibrational modes is called *vibronic coupling* and means that d–d transitions are observed, but for octahedral complexes they are often not very intense absorptions, and the observed colours are not very strong. In complexes of lower overall symmetry, for example tetrahedral molecules which have no centre of inversion, electronic transitions are more intense.

A familiar example of this effect is seen in the change of colour from deep blue to pale pink of self-indicating silica gel as if becomes damp. The intense blue colour arises from a small amount of a tetrahedral Co(II) complex added to the gel. As the gel absorbs water an octahedral Co(II) hexaaquo complex results. The change in colour reflects the change in energy separation on going from a tetrahedral to an octahedral geometry.

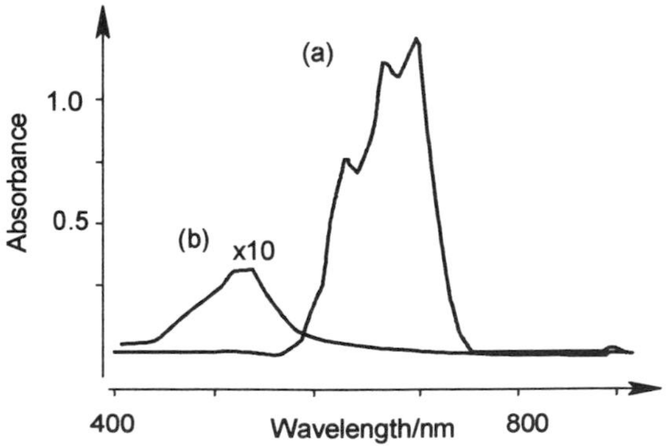

Fig. 4.19 Visible spectra for (a) tetrahedral $[CoCl_4]^{2-}$ and (b) octahedral $[Co(H_2O)_6]^{2+}$ complexes.

Since Δ_t is smaller than Δ_o the d–d transition in the octahedral complex will occur at a higher energy, and hence at a shorter wavelength. This blue-shifting of the absorption results in the appearance of a pink colour. The change in intensity arises from the transition being Laporte allowed for the tetrahedral complex where it is not for the octahedral complex. Figure 4.19 shows the visible spectra for these two complexes.

For high-spin octahedral d^5 complexes all d–d transitions are formally forbidden by all three selection rules. They are forbidden since $\Delta l \neq \pm 1$, they are Laporte forbidden, and *any* electronic transition must result in a change in the spin multiplicity of the complex as shown in Fig. 4.20 and therefore they are spin forbidden. For these reasons complexes containing high-spin d^5 transition metal ions such as Mn^{2+}-containing complexes tend to be very weakly coloured. The electronic spectra of such species, as shown in Fig. 4.21, frequently exhibit a relatively large number of weak electronic transitions corresponding to forbidden transitions.

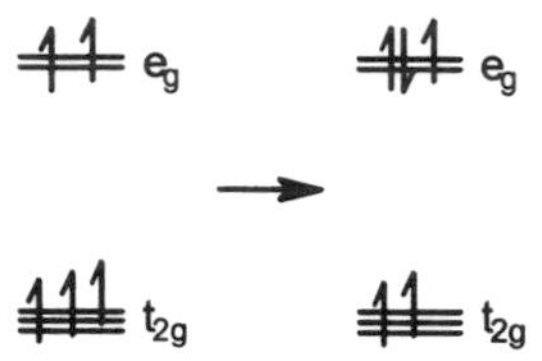

Fig. 4.20 Any d–d transition in a high-spin d^5 complex will result in a change in spin multiplicity.

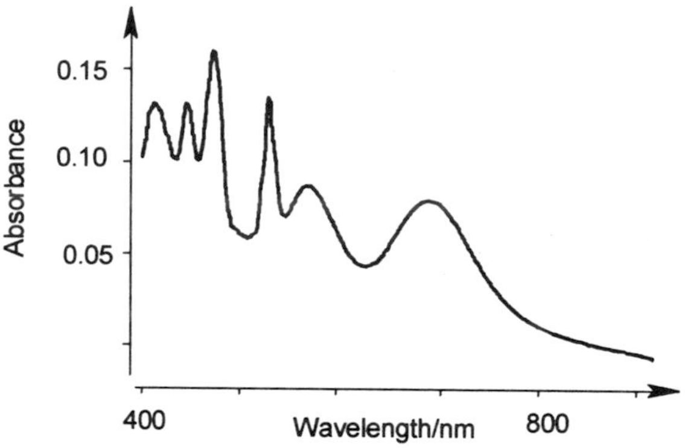

Fig. 4.21 The UV–visible spectrum of an aqueous solution containing $[Mn(H_2O)_6]^{2+}$.

Question 4.6 Which selection rules formally forbid d–d transitions in square-planar complexes?

We can use the selection rules to predict the relative intensities of the d–d transitions of compounds. For example, what order should be expected for the intensity of the d–d transitions of (i) $[MCl_6]^{2-}$, (ii) *trans*-$[M(H_2O)_4Cl_2]$, and (iii) *cis*-$[M(H_2O)_4Cl_2]$?

The principal difference that is going to affect the peak intensities between these compounds is the local geometry. In all cases the d–d transitions are formally disallowed. For the hexachloride the symmetry is octahedral and a centre of inversion exists which means transitions are additionally parity forbidden. In the *trans*-tetraaquodichloride complex the symmetry is no longer strictly octahedral but a centre of inversion still exists, whilst the *cis*-isomer does not possesses a centre of inversion. On this basis it would be expected that the d–d transitions should be most intense in complex (iii), then (ii), and weakest for the hexachloride (i).

Some typical extinction coefficient ranges for different types of electronic transitions are given in Table 4.4.

Electronic transition	ε_{max} ($l\ mol^{-1}\ cm^{-1}$)
Spin forbidden d–d	<1
Laporte forbidden d–d	20–100
Laporte allowed d–d	200–250
Charge transfer (symmetry allowed)	1000–50000

Table 4.4 Typical extinction coefficient values for some electronic transitions.

Jahn–Teller distortions

In a number of cases it has been observed that the metal–ligand bond lengths in a transition metal complex are not all equal as might be expected. For example, in certain electronic configurations of octahedral complexes, the two *trans*-bonds are measurably longer or shorter than the remaining four. The theory proposed by Jahn and Teller explains that when a set of degenerate orbitals (such as the e_g or t_{2g} orbitals of an octahedral complex) are unevenly filled, for example $(t_{2g})^1$ or $(e_g)^1$, then a perturbation of the molecular structure will occur so that the degeneracy is removed. Obviously, the greatest influence on the observed structure will arise when this additional splitting occurs to the orbitals lying along the same axes as the ligands, i.e. the e_g set of orbitals. Therefore, complexes possessing the configuration $(e_g)^1$ or $(e_g)^3$ are particularly susceptible to this phenomenon.

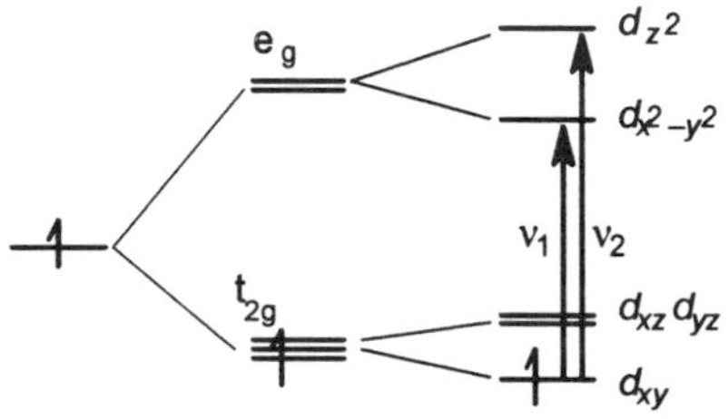

Fig. 4.22 The energy levels for $[Ti(H_2O)_6]^{3+}$ after Jahn–Teller distortion and the electronic transitions possible.

On this basis the $[Ti(H_2O)_6]^{3+}$ complex, which contains the Ti(III) ion and has a d^1 $[(t_{2g})^1\ (e_g)^0]$ ground state electronic configuration, is subject to a Jahn–Teller distortion resulting in a splitting of both the t_{2g} and e_g sets of orbitals, as shown in Fig. 4.22. If you look at the spectrum shown in Fig. 4.10(a) again you will see that there is a certain degree of asymmetry of the peak which may be due to this effect. In general, the reduction in the symmetry of a complex due to a Jahn–Teller distortion will result in a further splitting of the predicted energy levels and hence the potential for additional absorption bands in their UV–visible spectra. In extreme cases new bands will be observed, whilst in others a splitting of peaks or an asymmetry of some peaks may be observed.

Spectrochemical series

From early electronic spectroscopy studies it was noted that the positions of the absorption bands vary in a consistent manner for a set of metal ions and ligands. In particular, the ligands could be ordered in a series such that those higher in the list result in electronic transitions to higher energy. This ordered set of ligands is known as the *spectrochemical series.* Ligands high in this series result in large Δ_o on coordination to a metal.

$$I^- < Br^- < Cl^- < F^- < O^{2-} < OH^- < H_2O < NH_3 < NO_2^- < CN^- < PR_3 < CO$$

There are a number of different trends resulting in the final series. For the halide ligand the smaller, more electronegative ligands result in a greater splitting. The ability of a ligand to participate in π-bonding is also important. In general, π-donor ligands, such as halide ions, by virtue of their filled valence shells, lie at the low end of the series. Conversely, ligands which possess empty molecular orbitals of suitable energy so that they can act as π-acceptors, such as the cyanide ligand, carbon monoxide, and phosphines, are found at the high end of the series.

The size of the gap Δ_o will obviously have an effect on the frequency of light absorbed, the larger the gap the higher the energy of the electronic transitions. It will also have an important bearing on the arrangement of electrons within these levels. For ligands which result in a very large splitting of the t_{2g} and e_g sets of orbitals, low-spin configurations are expected to dominate, and vice versa.

The complex $[Co(NH_3)_5(NO_2)]^{2+}$ is yellow in colour whilst the linkage isomer $[Co(NH_3)_5(ONO)]^{2+}$ is red. What does this suggest about the position of the NO_2^- and ONO^- ligands in the spectrochemical series? The nitro-containing complex which appears yellow must be absorbing in the violet part of the visible spectrum and the nitrito-complex must be absorbing in the red part of the spectrum. Since violet light is of a higher energy than red, the energy gap Δ_o must be larger in the nitro-complex and therefore NO_2^- lies higher than ONO^- in the spectrochemical series.

	Δ_o (cm^{-1})		Δ_o (cm^{-1})
Ti^{3+}	20400	Fe^{3+}	21000
V^{3+}	19000	Fe^{2+}	10500
Cr^{3+}	17700	Co^{3+}	19000
Cr^{2+}	12500	Co^{2+}	9750
Mn^{3+}	21000	Ni^{2+}	8500
Mn^{2+}	7500	Cu^{2+}	12600

Table 4.5 Δ_o values for octahedral $[M(H_2O)_6]^{n+}$ complexes.

The energy gap will also be affected by the metal ion present in the complex. Table 4.5 gives some values for Δ_o for the hexaaquo complexes $[M(H_2O)_6]^{n+}$. Generally, a trend is observed such that the separation between the t_{2g} and e_g orbitals decreases for the heavier metal. The lower oxidation state complexes for the same metal have a considerably smaller energy gap because in the electrostatic bonding model the higher the charge on the metal the stronger the electrostatic attraction. The more highly oxidised metal will also possess a smaller radius. Both of these factors result in a stronger interaction between ligands and metal and so a larger Δ_o value.

Data are available from a number of different sources for Δ_o for a variety of different ligands which may be used to approximate the energy gap in other related systems. For example. Δ_o for $[Co(NH_3)_6]^{3+}$ is 22900 cm^{-1}, and for $[Co(H_2O)_6]^{3+}$ 19000 cm^{-1}, we can use a weighted average of these two values to obtain an estimate for Δ_o for the intermediate complexes, such as $[Co(NH_3)_3(H_2O)_3]^{3+}$

$$\Delta_o = \frac{3}{6} \times 19000 + \frac{3}{6} \times 22900 \text{ cm}^{-1}$$
$$= 20950 \text{ cm}^{-1}$$

Question 4.7 Δ_o for $[Rh(en)_3]^{3+}$ is 32700 cm^{-1}, and 19300 cm^{-1} for $[RhCl_6]^{3-}$. For an octahedral complex $[Rh(en)_xCl_y]^{n+}$ Δ_o is 28400 cm^{-1}: what are the most likely values of *x* and *y*?

However, do not expect these values to be particularly accurate (one reason being that the symmetry of the intermediate complex is no longer strictly octahedral).

4.4 Charge-transfer transitions

There are some transition metal complexes that are intensely coloured but which possess no d-electrons, for example potassium permanganate gives a very intensely purple solution when dissolved in water yet the permanganate ion (MnO_4^-) is formally an Mn(VII) species and therefore d^0. Similarly, there are a number of coloured main group compounds such as HgS which also do not have partially filled d-orbitals. The colours of these compounds still arise from electronic transitions, but now from a change in the electron distribution between the metal and the ligands. These are called *charge transfer* (CT) bands. Such transitions are formally allowed by the selection rules and are therefore much more intense (frequently by a factor of a thousand or more) compared with the forbidden d–d transitions.

Metal–ligand transitions

All ligands will possess a number of molecular orbitals which may be of σ-, σ*-, π-, π*- or non-bonding (n) character. If these orbitals are filled and those on the metal are empty then charge transfer may occur from the filled ligand-based orbitals to the empty metal d-orbitals. Absorptions arising from such processes are called *ligand-to-metal* charge-transfer bands and abbreviated to LMCT. If the ligand possesses low-lying empty orbitals such as CO or CN^- and the metal is in a low oxidation state and hence electron rich, then *metal-to-ligand* charge transfer (MLCT) transitions will occur.

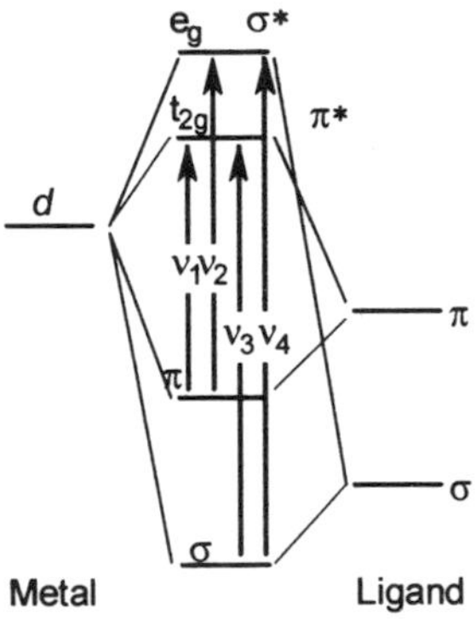

Fig. 4.23 Simplified MO diagram showing the orbitals involved in L→M charge-transfer transitions.

Figure 4.23 shows a simplified molecular orbital diagram for ligand σ- and π-orbitals and the metal t_{2g} and e_g orbitals. From these, four transitions are possible and these are labelled on the diagram. As the figure is drawn these transitions should occur in the following order of increasing energy: $\pi \rightarrow t_{2g} < \pi \rightarrow e_g < \sigma \rightarrow t_{2g} < \sigma \rightarrow e_g$. For 3d transition metals such transitions usually occur at much higher energy than d–d absorptions, lying at the extreme blue end of the visible spectrum or in the UV (<400 nm). However, for complexes of the 4d and 5d transition metals these bands are at lower energy and frequently some of the weaker d–d transitions are swamped by charge-transfer bands. The transitions marked as ν_3 and ν_4 are often broad and beyond the instrument range. Of course, for transition metal complexes with filled t_{2g} orbitals, transition ν_1 is not possible.

In the case of the permanganate ion which contains Mn(VII) the metal d-orbitals are empty and therefore the intense colour must arise from a ligand-to-metal charge transfer process. There are routes available for calculating the position of the lowest charge transfer band for a particular ligand and metal centre given below; however, a guide to the relative positions of these bands may be obtained by considering the charge-transfer process as a form of redox process. For a ligand-to-metal charge transfer the ligand is formally oxidised and the metal reduced and vice versa for a metal-to-ligand charge transfer.

We should expect that in systems containing high oxidation state (and hence electron-poor) metals such as the hexahalides MF_6, MCl_6, and MBr_6, charge transfer is most likely to occur from the ligand to the metal. Since bromide is more easily oxidised than chloride or fluoride the first charge-transfer band for the hexabromide will be at lower energy than the hexachloride and then the hexafluoride. Similarly, for the $[MCl_6]^{2-}$ and $[MCl_6]^{3-}$ ions which contain M(II) and M(III) centres respectively it should be easier to oxidise the M(II) centre than the M(III) centre and therefore the lowest energy metal–ligand charge transfer band for $[MCl_6]^{2-}$ should be at a lower energy than for $[MCl_6]^{3-}$. Values for the positions of some charge transfer transitions are given in Table 4.6.

Complex	Lowest energy charge transfer band (cm^{-1})
$[RuCl_6]^{2-}$	17150 ($\pi \rightarrow t_{2g}$)
$[RuCl_6]^{3-}$	25600 ($\pi \rightarrow t_{2g}$)
$[OsCl_6]^{3-}$	23900 ($\pi \rightarrow t_{2g}$)
$[OsBr_6]^{2-}$	21000 ($\pi \rightarrow t_{2g}$)
$[OsI_6]^{2-}$	15000 ($\pi \rightarrow t_{2g}$)
$Mo(CO)_6$	35600 ($M \rightarrow L\pi^*$)
$[MnO_4]^-$	18900 ($n \rightarrow \pi^*$)

Table 4.6 Charge transfer bands for some transition metal complexes.

Question 4.8 Would you expect the lowest energy charge-transfer band of $[ReO_4]^-$ to be above or below that of $[MnO_4]^-$?

Optical electronegativity

There is a method available for predicting the position of the lowest charge-transfer band for a complex based on its constituent metal and ligands. This method was proposed by Jørgensen and takes the form shown below

$$\bar{\nu} = 30000 \times (\chi^{X}_{opt} - \chi^{M}_{opt})\ cm^{-1} \qquad (4.4)$$

where χ^{X}_{opt} and χ^{M}_{opt} are the optical electronegativity values of the ligand the metal respectively. Some typical values are given in Table 4.7.

Ligand	χ^{X}_{opt}	Metal	χ^{M}_{opt}
F^-	3.9	Cr(III)	1.9
Cl^-	3.0	Mn(III)	2.0
Br^-	2.8	Fe(III)	2.1
I^-	2.5	Co(III)	2.3
CN^-	3.5	Ni(II)	2.1
H_2O	2.5	Cu(II)	2.3

Table 4.7 Some optical electronegativity values for some ligands and octahedrally coordinated metal ions.

4.5 Ligand-centred transitions

Of course there need not be a metal involved in an electronic transition at all. Ligands, particularly those containing π–bonds, will have their own electronic spectra arising from promotion of electrons from the filled bonding molecular orbitals to anti-bonding or non-bonding levels. These systems are often referred to as *chromophores*. Figure 4.24 shows the types of transitions that are possible for a single chromophore, and how, as more π-systems are linked together or conjugated, the energy gap lessens and so electronic transitions are observed at lower energy or longer wavelength values. As the degree of conjugation of double bonds increases, the molecular diagram will possess more bonding and anti-bonding orbitals between which transitions may occur.

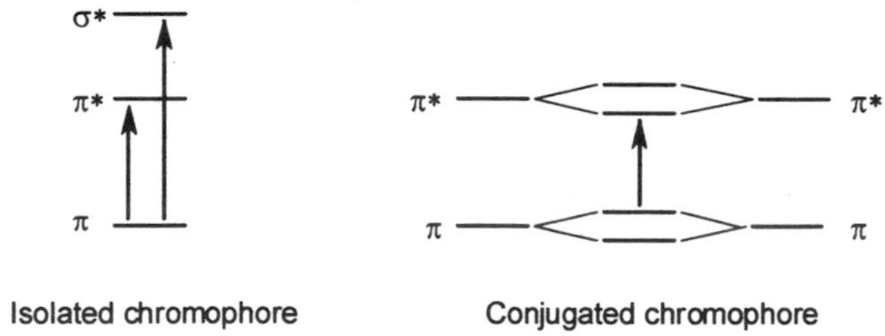

Fig 4.24 Generic MO diagrams for single and conjugated chromophores.

Typical absorption maxima for some common organic groups and ligands are given in Table 4.8. From the first few entries it is seen that as the degree of conjugation between double bonds increases, the maximum absorption peak shifts to lower energy and hence higher wavelengths. Certain other inorganic ligands such as thiocyanide, nitrate, and nitrite also possess very intense charge transfer bands around 200 nm. Because these transitions arise from molecular orbitals within the ligand, coordination to metals normally has only a small effect on the position of these absorptions.

Compound	λ_{max} (nm) [ε (l mol^{-1} cm^{-1})]
>C=C<	180–200 [10000]
>C=CH–CH=C<	210–220 [20000]
C_6H_6	184 [60000]
PPh_3	215 [29000]
PMe_3	201 [18800]
$P(OMe)_3$	190 [63100]
$AsPh_3$	248 [13200]

Table 4.8 The lowest energy charge-transfer bands for some ligands and organic molecules.

4.6 Possible problems

The following section outlines some of the potential problems in assigning UV–visible spectra and some possible solutions.

Problem	Possible cause	Suggested remedy
More peaks than expected	(a) Mixture of compounds (b) Lower molecular symmetry than expected (c) Jahn–Teller distortion	(a) Effect separation if possible
Fewer peaks than expected	(a) Molecule possesses higher symmetry than expected (b) Accidental overlap of peaks, e.g. CT bands obscuring d–d transitions	(a) Record spectra for a more concentrated sample to ensure very weak peaks are not missed
Many very weak sharp peaks	d^5 metal ion	

4.7 Further questions

4.9 The UV–visible spectrum of RuF_6 exhibits the following absorptions: 18400(vw), 23680(w), 25600(w,sh), 32800(m), 37940(m), 51000(s) cm^{-1}. Assign these bands as far as possible and make an estimate of χ_{opt}(Ru(VI)).

4.10 The UV–visible spectrum in Fig. 4.25 is from a solution of 0.12 g $CrCl_3$ in 25 cm^3 of water. Δ_o for the complex $[CrCl_6]^{3-}$ is 13800 cm^{-1} and for $[Cr(H_2O)_6]^{3+}$ is 17400 cm^{-1}. Assign the bands where possible and hence estimate Δ_o for the solution-phase species and determine its composition.

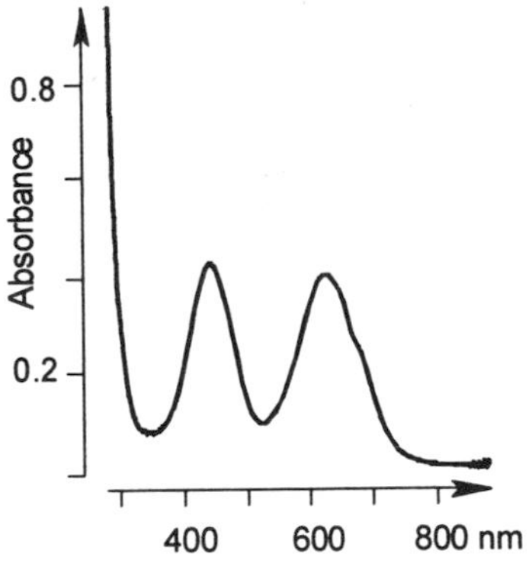

Fig. 4.25 The UV–visible spectrum of $CrCl_3$ dissolved in water.

Further reading

M. Gerloch and E.C. Constable, Transition metal chemistry, VCH (1994).
A.B.P. Lever, Inorganic electronic spectroscopy, Elsevier (1968).
C.K. Jørgensen, Absorption spectra and chemical bonding in complexes, Pergamon Press (1964).

5 Mass spectrometry

5.1 Introduction

Although mass spectrometry is not strictly a form of spectroscopy, in that it does not involve the interaction of radiation with matter, it is none the less a very useful method for identifying unknown compounds. One of the reasons it is so useful is that it can provide information about the relative molecular mass of a compound. Once we know this we can compare it with that calculated for the expected products based on the relative atomic masses of the constituent elements and the composition of the molecule (i.e. its molecular formula). In fact, mass spectroscopy will usually provide us with a lot more information than just the mass of compounds; it is frequently possible to determine which ligands or groups of atoms are bonded together from the way the compound breaks up or *fragments* during the experiment.

5.2 The basics

Mass spectroscopy requires that the molecule of interest is charged; this then provides a basis for separating the different ions due to their mass-to-charge ratio (m/z or m/e) and this can be achieved in a number of different ways. Figure 5.1 shows a schematic representation of the processes involved in mass spectrometry.

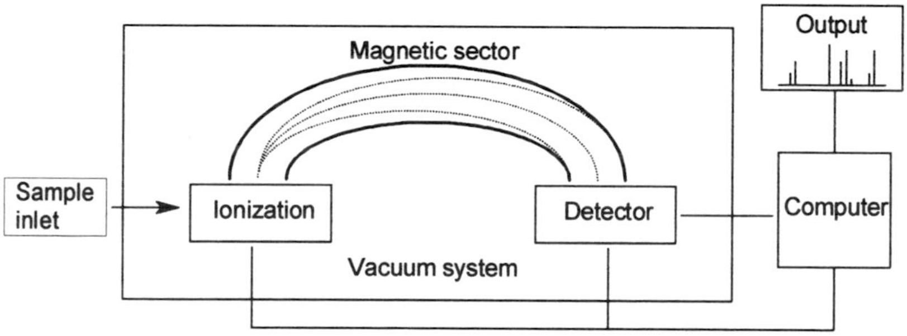

Fig. 5.1 A schematic of a mass spectrometer.

There are a number of different methods of separating charged ions; one of these relies on a magnetic field to deflect the particle. If an ion, of mass m and charge z, is accelerated by a fixed voltage, V, then each molecule will acquire a velocity, v, and kinetic energy given by

$$zV = \tfrac{1}{2}mv^2 \qquad (5.1)$$

When these ions are subjected to a magnetic field, B, they will undergo a curved trajectory of radius r according to

$$Bzv = mv^2/r \tag{5.2}$$

Combining eqns 5.1 and 5.2 results in an expression that relates the applied magnetic field to the mass to charge ratio

$$\frac{m}{z} = \frac{B^2 r^2}{2V} \tag{5.3}$$

Thus by varying the magnetic field it is possible to select which ions undergo sufficient curvature of their path so that they reach a detector. The spectrum produced by a mass spectrometer is, therefore, a plot of the number of ions detected versus their mass-to-charge ratio, m/z. The unit of mass most commonly used in mass spectrometry is the atomic mass unit or dalton (Da) which is defined as one-twelfth of the mass of a carbon-12 atom.

There are other methods of separating ions, one of which, called time-of-flight, provides all molecules with a certain amount of momentum and then allows them to drift towards the detector; the rate at which this occurs will vary depending on the mass of the ions. Another common method of selecting ions is called quadrupole detection and is based on alternating electric fields applied across four poles (hence the name). By varying the frequency of these electric fields, ions with different mass-to-charge ratios will pass between the poles and reach the detector at the other end. Irrespective of the methods used, mass spectrometers do not directly provide mass measurements but separate ions on the basis of their mass-to-charge ratio.

5.3 Ionisation methods

As mentioned above, molecules need to be ionised before they can be selected by their mass-to-charge ratio. The method used to obtain these ions often has a significant bearing on the final spectrum obtained and is dealt with in more detail below. It is, of course, important that the ions reach the detector without combining or reacting and therefore a mass spectrometer chamber is held under high vacuum so that the mean-free path of the ions before collision is long enough that the majority reach the detector.

Electron impact

One of the most common methods of ionisation is electron impact (EI), where a beam of accelerated electrons, usually derived from a heated tungsten filament, bombard molecules causing ionisation. The ionised molecules are then accelerated into the detection part of the mass spectrometer.

$$M + e^- \rightarrow M^{+\cdot} + 2e^-$$

Ideally, molecules will just be singly ionised, but it is possible for multiple ionisations to occur.

$$M + e^- \rightarrow M^{n+\cdot} + (n+1)e^-$$

When this does arise under normal conditions it is limited to double or triple ionisation, although one method of measuring the mass of very large molecules is to deliberately ensure that molecules are highly charged so they appear at lower *m/z* values.

Fast atom bombardment

There are occasions when heating a sample and bombarding it with accelerated electrons provides too much energy and the molecules are not just ionised but also start to break up into smaller *fragments*. This is especially true for molecules containing relatively weak bonds or easily fragmented groups such as organometallic compounds; in these cases the technique of fast atom bombardment (FAB) is often preferred. In this method the compound of interest is dissolved in a suitable non-volatile 'matrix' material such as glycerol, before being bombarded by charged atoms of argon or xenon. This method results in a higher chance of observing the *parent* or *molecular ion* with less fragmentation; however, it may also give rise to peaks due to the matrix material or even peaks arising from the products of reaction between the matrix and compound.

Chemical ionisation

Another common ionisation method is called chemical ionisation. In this case gas molecules and ions collide, resulting in the generation of gaseous ions with low kinetic energy (AH^+) which are then used to ionise other neutral molecules. Gases such as methane and ammonia are commonly used for this technique giving rise to the ionising species CH_5^+ and NH_4^+.

$$AH^+ + M \rightarrow A + MH^+$$

Under such ionisation conditions peaks are observed in the mass spectrum corresponding to the molecular ion plus H (i.e. *m/z* M+1) and the molecular ion plus the protonated ionising gaseous molecule $[M+AH]^+$ (i.e. *m/z* = M+17 for CH_5^+ or M+18 for NH_4^+).

5.4 Interpretation of mass spectra

Having obtained an ionised sample, a number of different mass spectrometric experiments are available to identify compounds, some of which are described below.

Accurate mass measurements

It is possible for the mass-to-charge ratio of ions to be determined to one part in a million, or better. This may not appear at first glance to be of great use; however, consider an attempt to differentiate between CO and N_2 gases using mass spectrometry. Both have a mass of 28 amu based on the masses of the elements, C = 12, N = 14, and O = 16, and it would therefore appear that they cannot be distinguished using this method. However, when we use the accurate masses of the elements, given in Table 5.1, we find that the relative molecular mass (rmm) of CO is 27.9994 and of N_2 is 28.01348 and thus these two molecules may be readily identified by a high-resolution mass spectrometry study.

Element	Accurate mass
C	12.0000
N	14.00674
O	15.9994

Table 5.1 The accurate relative atomic masses of carbon, nitrogen, and oxygen.

Isotope patterns

We do not need to record extremely accurate mass-to-charge ratios to determine the identity of many inorganic compounds, since many elements naturally possess a number of different isotopes which give rise to identifiable isotope patterns. There are also a number of the common elements which are either mono-isotopic or have essentially only one naturally occurring isotope. Some of the elements that fall into this category are given in Table 5.2. Any molecule that contains isotopic elements will exhibit a number of peaks in its mass spectrum, the relative intensities of which reflect the isotopic abundances.

H, Be, C, N, F, Na, Al, P, Sc, Mn, Co, As, Y, Nb, Rh, I, Cs, Ta, Au, Bi

Table 5.2 The elements that are mono-isotopic, or may be considered as such by virtue of possessing one very high abundance isotope.

These characteristic patterns arising from the natural relative abundance of the isotopes (so called isotope patterns) are shown in Fig. 5.2 for a number of the more common elements.

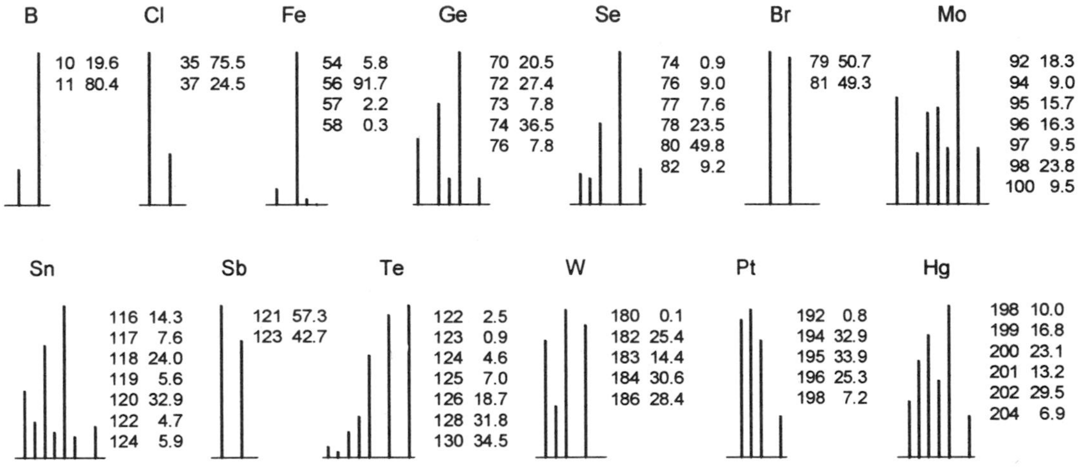

Fig. 5.2 Natural abundance of some of the more common elements.

Bromine has two naturally occurring isotopes, ^{79}Br which is 50.7% abundant and ^{81}Br (49.3% abundant), as shown in Fig. 5.2. On average, slightly over half of all the molecules of a particular compound containing a single bromine atom will be composed of ^{79}Br and the remainder ^{81}Br. So the mass spectrum of the *molecular ion* of HBr would consist of two lines at m/z = 80 [1 (H) + 79 (Br)] corresponding to H^{79}Br and m/z = 82 (due to H^{81}Br). The intensity of these two lines will reflect their natural abundance, i.e. 50.7%:49.3%, or roughly 1:1. So the mass spectrum of the HBr ion should appear as two lines of similar intensity at m/z values of 80 and 82, as shown in Fig. 5.3. The same intensity pattern will be observed for *any* molecule which contains one bromine atom and other elements which are mono-isotopic, for example PH_2Br and CF_3Br.

What if there were two bromine atoms in a molecule, such as $PFBr_2$? We now need to consider a larger number of possibilities; both phosphorus and fluorine are mono-isotopic but some molecules will contain two ^{79}Br atoms, some two ^{81}Br atoms, and some a mixture of one ^{79}Br atom and one ^{81}Br atom. We would therefore expect to see peaks in the mass spectrum corresponding to *m/z* values of 208 ($PF^{79}Br^{79}Br$), 210 ($PF^{79}Br^{81}Br$), and 212 ($PF^{81}Br^{81}Br$). The relative intensities of these lines will be based on the natural abundance of the bromine isotopes contained in each molecule. Thus the molecules with two ^{79}Br atoms should have a relative abundance of 50.7% × 50.7% (i.e. $0.507^2 = 0.257$), and the molecules containing two ^{81}Br atoms should have a relative abundance of $0.493^2 = 0.243$. The remainder of the molecules, by definition, must contain both ^{79}Br and ^{81}Br but there is a choice of which of the bromine atoms is ^{79}Br or ^{81}Br. Once one is chosen the other must be fixed so there are two possible permutations, $PF^{79}Br^{81}Br$ and $PF^{81}Br^{79}Br$. Since both of these result in the same mass, both peaks will occur in the same place in the mass spectrum and thus the total intensity of this peak will be $(0.507 \times 0.493) + (0.493 \times 0.507) = 0.500$.

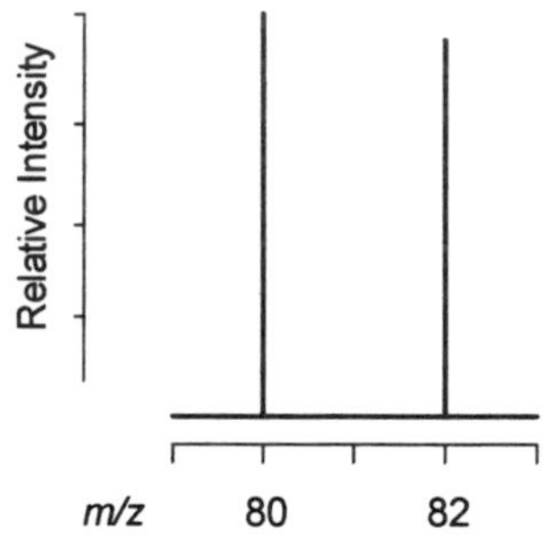

Fig. 5.3 Calculated mass spectrum for HBr.

A more long-winded version, especially when there are many different isotopic elements in a molecule, is to write out all the possible isotopic species and calculate their individual masses and abundances and then sum all the abundances for similar mass peaks.

Molecule	Mass	Abundance
$PF^{79}Br^{79}Br$	208	0.507 × 0.507 = 0.257
$PF^{79}Br^{81}Br$	210	0.507 × 0.493 = 0.250
$PF^{81}Br^{79}Br$	210	0.493 × 0.507 = 0.250
$PF^{81}Br^{81}Br$	212	0.493 × 0.493 = 0.243
		Total = 1.000

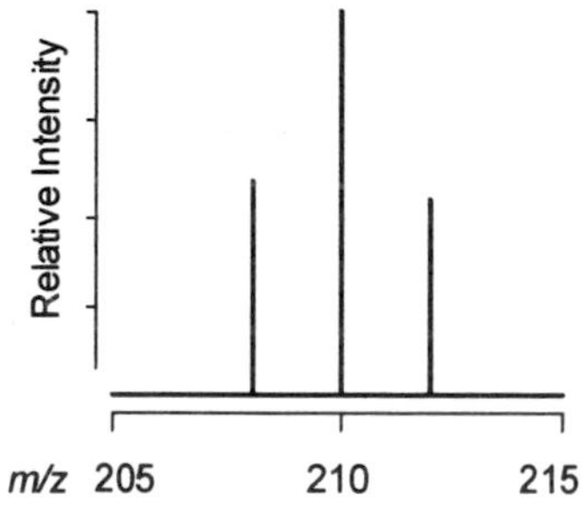

Fig. 5.4 Isotopic peaks for the molecular ion of $PFBr_2$.

Provided all possibilities have been included, and the calculations are correct, then the sum of all the abundances should add up to unity (1.0), and this may be used as a check. Therefore, we would predict that the mass spectrum for the $PFBr_2$ ion would consist of three peaks at *m/z* = 208, 210, and 212 in the relative ratio of 0.257:0.500:0.243 (or roughly 1:2:1). The mass spectrum of the molecular ion is shown in Fig. 5.4

Two isotopomers are two compounds which differ only in the isotopic mass of one or more of the elements involved.

What would the isotopic distribution pattern for PCl_3 be? The natural abundance of chlorine is ^{35}Cl 75.8%, ^{37}Cl 24.2%, which are almost in the ratio 3:1 and this will be used since it simplifies the calculations. Isotopomers of PCl_3 may contain either (i) three ^{35}Cl, or (ii) two ^{35}Cl and one ^{37}Cl, or (iii) one ^{35}Cl and two ^{37}Cl, or (iv) three ^{37}Cl atoms. Such arrangements can be met in only one way for (i) and (iv), but there will be three combinations for both (ii) and (iii). The relative intensities will therefore be the relative abundance for each chlorine atom multiplied by the number of possible permutations. Taking each isotopomer in turn this gives (i) 3×3×3×1, (ii) 3×3×1×1, (iii) 3×1×1×1, and (iv) 1×1×1×1. We therefore see four peaks, each separated by 2 mass units in the ratio 27:9:6:1.

The isotope patterns expected for a molecule containing mono-isotopic elements and a certain number of chlorine or bromine atoms can be quite indicative as shown by the patterns in Fig. 5.5 which are for one to six of these halogen atoms.

Question 5.1 Calculate the isotope pattern for the ion $BrCl^+$.

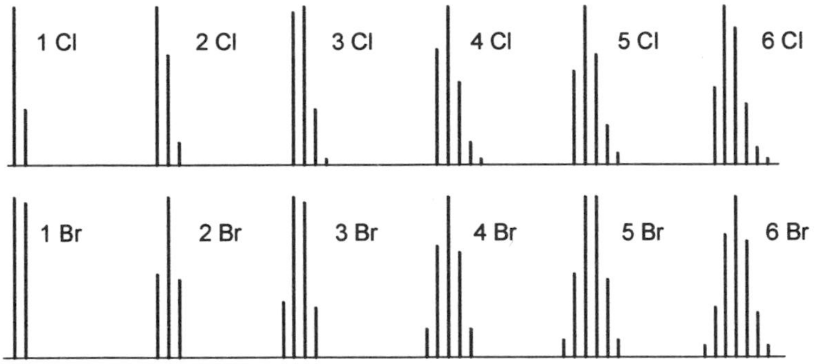

Fig. 5.5 Calculated isotope patterns for a number of chlorine and bromine atoms.

Fragmentation patterns

So far we have concentrated on the appearance of the *molecular* or *parent ion* peaks. However, it is usual under the conditions of the experiment for fragmentation of the molecule to occur and this will result in the observation of *fragment ions*. Although this might be considered an undesired side-effect it can often be very informative.

The mass spectrum shown in Fig. 5.6 is of carbon dioxide which has been ionised by the electron impact method. The most intense peak is due to the molecular ion, normally indicated by the notation M^+, and appears at m/z of 44 [12 (C) + 2 × 16 (O)]. However, weaker peaks are also observed at m/z values of 28, 16, and 12. These arise from fragmentation of the parent molecule and correspond to CO^+, O^+, and C^+ ions, respectively.

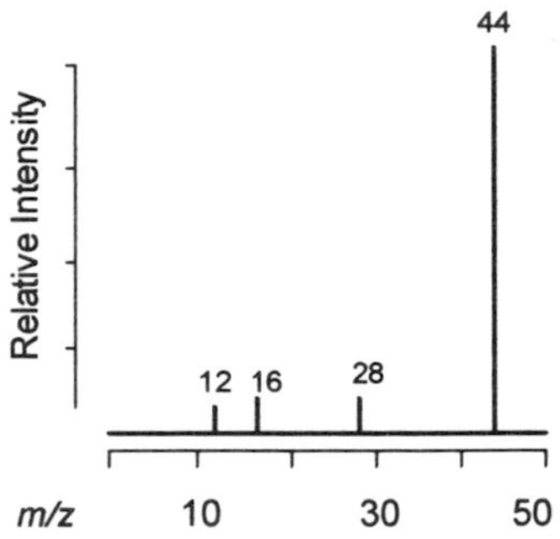

Fig. 5.6 EI spectrum of CO_2.

The appearance of ions due to fragmentation can allow us to distinguish between other possible molecules with a similar mass. For example, CO and N_2 both have an rmm of 28 amu and it has already been shown that a high-resolution study can be used to distinguish between these two, although in a low-resolution mass spectrometric study both will exhibit a parent ion peak at $m/z = 28$. For dinitrogen the only possible fragment which would arise from breaking the N–N bond is the N^+ ion with m/z of 14. However, for carbon monoxide we expect to observe peaks due to the fragments C^+ ($m/z = 12$) and O^+ ($m/z = 16$). It should therefore be possible to distinguish between dinitrogen and carbon monoxide either using an accurate mass experiment or by studying their fragmentation patterns.

For larger molecules there are frequently a number of ways a molecule may fragment

$$\begin{aligned} ABC + e^- &\rightarrow AB^+ + C + 2e^- \\ &\rightarrow BC^+ + A + 2e^- \end{aligned}$$

and the fragments themselves may fragment further

$$AB^+ \rightarrow A^+ + B$$
$$\rightarrow A + B^+$$

Such processes frequently occur for compounds containing a number of relatively weak bonds, for example in organometallic compounds.

Question 5.2 What possible fragments may arise from PMe_3?

The mass spectrum of an iron carbonyl is shown in Fig. 5.7. The spectrum exhibits a number of groups of peaks centred at *m/z* values of 196, 168, 140, 112, 84, and 56. In this case the pattern for each group of peaks appears to be similar and corresponds to that due to the natural isotopic abundance of iron, shown in Fig 5.2. The lowest mass set of peaks is at *m/z* of 56 and since this corresponds to the highest natural abundance isotope of iron we can assign this to the Fe^+ fragment. The most intense peak in each set of peaks is separated from the next by 28 amu, suggesting that a number of fragments of the same mass are being lost many times. In this case this is most likely to be carbon monoxide (28 amu). In total, the spectrum exhibits six sets of peaks each separated by 28 units and displaying the iron isotope pattern; we can therefore assign these to the ions $Fe(CO)_x^+$ ($x = 0$ to 5). The mass spectrum, therefore, suggests that $Fe(CO)_5$ was the compound that produced this spectrum.

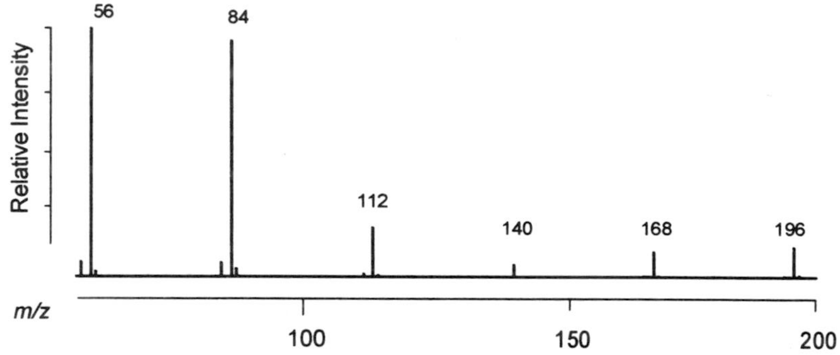

Fig. 5.7 The mass spectrum for an iron carbonyl compound.

We can sketch the fragmentation pathway that gives rise to these peaks as an aid to interpreting the mass spectrum. In this case, shown in Fig. 5.8, successive fragmentation of carbonyl ligands occurs to result in the observed spectrum.

$$Fe(CO)_5 \xrightarrow{-CO} Fe(CO)_4 \xrightarrow{-CO} Fe(CO)_3 \xrightarrow{-CO} Fe(CO)_2 \xrightarrow{-CO} Fe(CO) \xrightarrow{-CO} Fe$$

Fig. 5.8 Fragmentation pathway for $Fe(CO)_5$.

The appearance of fragmentation patterns can be used to identify unknown compounds since it allows the molecule to be 'put back together again'. For example, Fig. 5.9 shows the mass spectrum of a compound obtained by the reaction of $TiCl_4$ and Na^+Cp^- ($Cp^- = \eta^5$-$C_5H_5^-$).

The starting point in assigning this spectrum and hence identifying the compound produced is to determine the atomic masses and isotope distributions for the elements involved. Hydrogen and carbon can be considered as effectively mono-isotopic so any hydrocarbon fragment will appear essentially as a single peak. The remaining elements all have more than one isotope, chlorine has two naturally occurring isotopes (35 and 37 amu) in the approximate ratio 3:1. Although titanium has a total of five isotopes, the distribution is dominated by a single isotope of mass 48 as shown in the isotope pattern diagram in Fig. 5.10.

Question 5.3 The mass spectrum of SO_2 shows peaks at *m/z* = 16, 32, 48 and 64; what ions may these peaks correspond to?

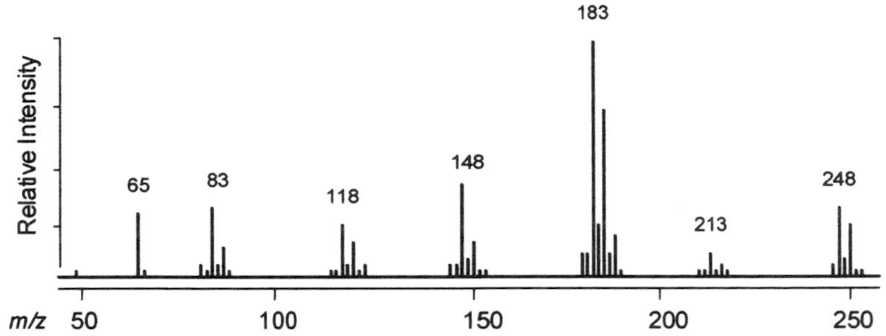

Fig. 5.9 The EI mass spectrum obtained for the product of reaction between $TiCl_4$ and Na^+Cp^- ($Cp^- = \eta^5$-$C_5H_5^-$).

So how do we proceed in assigning the peaks? It is possible that the very small peak just below 50 amu is due to Ti^+ but we cannot be sure since we cannot clearly see its isotope pattern. We therefore start by looking for patterns that we do recognise, either the titanium pattern or the characteristic chlorine isotope patterns. We also need to determine the separations between sets of peaks and try and tie these in with possible ligands and fragments. Looking at the spectrum we recognise that there appear to be two repeating patterns. The sets of peaks at *m/z* values of 83, 148, and 213 are dominated by a single intense peak but accompanied by a number of weaker ones, whilst the second set of peaks at *m/z* values of 118, 183, and 248 have two more intense peaks and a number of weaker ones in each group. The separation between each of the sets of peaks of similar appearance is 65 amu (213 – 148 = 65, 148 – 83 = 65, and 248 – 183 = 65, 183 – 118 = 65). The cyclopentadienyl ligand ($C_5H_5^-$) has an rmm of 65 suggesting that these differences might correspond to the successive loss of this ligand. The mass difference between the two different sets of peaks is 35 amu (248 – 213 = 35, 183 – 148 = 35, and 118 – 83 = 35) which corresponds to the rmm of the most abundant isotope of chlorine suggesting that at least one chloride ligand has fragmented from the compound.

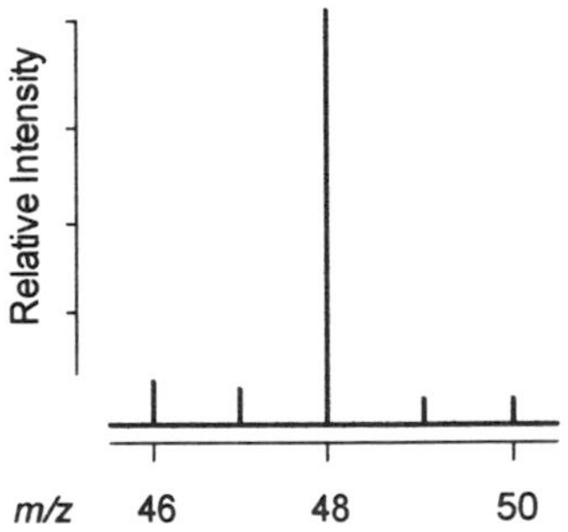

Fig. 5.10 The isotope pattern for titanium.

So putting together what we have so far, we can account for 48 (Ti) + 2×65 (two Cp ligands) + 35 (Cl) = 213. This does not correspond to the highest *m/z* peak at 248; the difference between these two figures is 35, i.e. another chloride ligand. The presence of a different number of chlorine atoms would account for the different appearance of the two sets of peaks. Figure 5.5 shows that the presence of two chlorine atoms in a molecule should produce a pattern with peaks in the approximate ratios of 9:6:1, like that seen on the peaks at 248, 183, and 118. When there is just one chlorine atom we would expect an isotope pattern that resembles more closely the 3:1 ratio of ^{35}Cl:^{37}Cl isotopic distribution. So we assume (in the absence of any indications otherwise) that the highest mass peaks correspond to the molecular ion—Cp_2TiCl_2. The lower-mass peaks therefore correspond to the fragment ions Cp_2TiCl^+ ($m/z = 213$), $CpTiCl_2^+$ ($m/z = 183$), $CpTiCl^+$ ($m/z = 148$), $TiCl_2^+$ ($m/z = 118$), $TiCl^+$ ($m/z = 83$) and the peak at 65 amu is due to the $C_5H_5^+$ fragment. This can be confirmed by calculating the isotopic masses and relative intensities for each of the fragments in turn.

Figure 5.11 shows the possible fragmentation pathways for this compound which would give rise to the identified fragments.

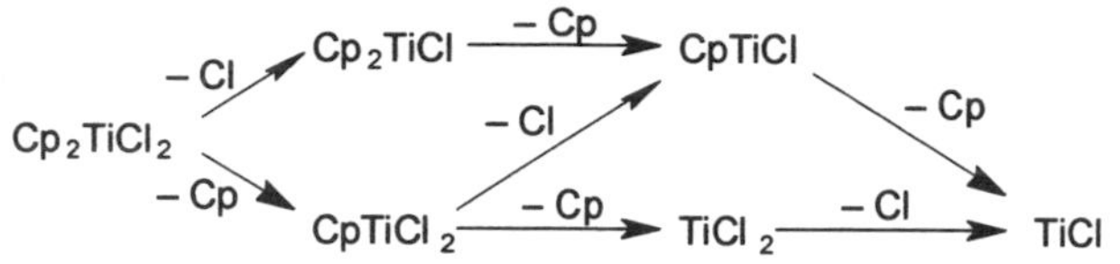

Fig. 5.11 Fragmentation pathways for Cp_2TiCl_2.

Problems may arise in trying to recognise isotope and fragmentation patterns if overlap occurs between two different patterns. This is especially likely to occur if hydrogen can be fragmented from a molecule, for example Fig. 5.12 shows the mass spectrum of B_2H_6. There are two clusters of peaks just below the expected masses for the parent ion $B_2H_6^+$, $m/z = 26$, and the fragment BH_3^+ ($m/z = 15$). However, neither of these patterns corresponds to the natural abundance of the two boron isotopes (^{10}B:^{11}B 3:1, as shown in Fig. 5.2). The presence of lower-mass peaks suggests that fragmentation has occurred which results in the overlap of peaks from a number of different ions. In the case of the lower-mass peaks this can be approximated to a 2:1:1.5 mixture of the ions BH_2^+:BH^+:B^+ when a good match to the observed spectrum is obtained.

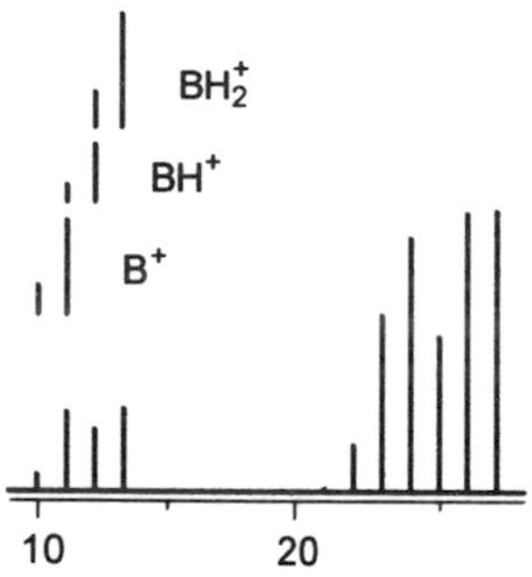

Fig 5.12 The observed and calculated mass spectra for B_2H_6.

Question 5.4 Why is it unlikely that the set of peaks in the above spectrum around $m/z = 10$ is due to $B_2H_6^{2+}$?

Rearrangements

It is quite common for mass spectra, especially those of organometallic and organic compounds, to show peaks which cannot be assigned to fragments of the compound under investigation, but can be assigned to completely different molecules. Such species can arise from rearrangement reactions which occur within the mass spectrometer. One possible rearrangement resulting in the production of two new molecules is shown in Fig. 5.13

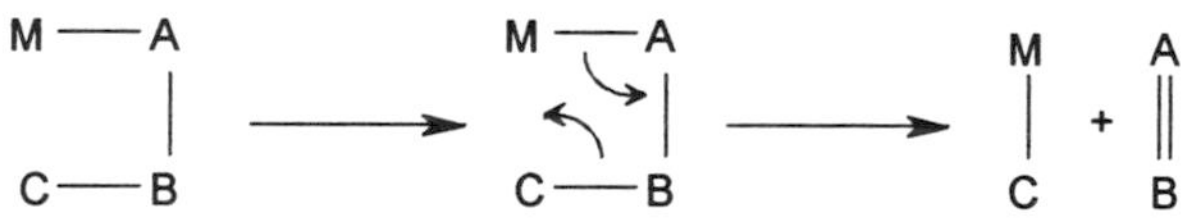

Fig. 5.13 Schematic representation of a molecular arrangement which can occur in a mass spectrometer.

There are many instances when such rearrangements occur. Metal complexes containing halogenated organic ligands frequently undergo rearrangements resulting in the formation of a metal–halogen bond by the mechanism outlined above. For example, the mass spectrum of $Si(C_6F_5)_4$ under electron impact conditions shows, amongst others, peaks corresponding to SiF^+, SiF_3^+, and $SiF(C_6F_5)^+$.

5.5 Possible problems

Some of the potential problems that may be associated with the analysis of compounds by mass spectrometry are listed below with suggested remedies.

Problem	Possible cause	Suggested remedy
No peak corresponding to the molecular ion observed	(a) Incorrect assignment of compound (b) Extensive fragmentation	(a) Use other spectroscopic methods to confirm, or otherwise, compound's identity (b) Use 'softer' ionisation method, e.g. FAB
Parent peak mass appears to be too high	(a) Incorrect assignment of compound (b) Reactions, e.g. with FAB matrix or CI gas (c) Peaks due to $[M+AH]^+$ for CI	(a) See above. (b) Try alternative ionisation methods
The mass distribution of a set of peaks is correct but the isotope pattern is not	The patterns from two or more fragments are overlapping	Use an alternative ionisation method to reduce fragmentation or use less energetic electrons for EI method.

5.6 Further questions

5.5 The EI spectrum, Fig. 5.14, is of $Fe(acac)_3$ [acac = $CH_3COCH_2COCH_3^-$]; assign the peaks where possible.

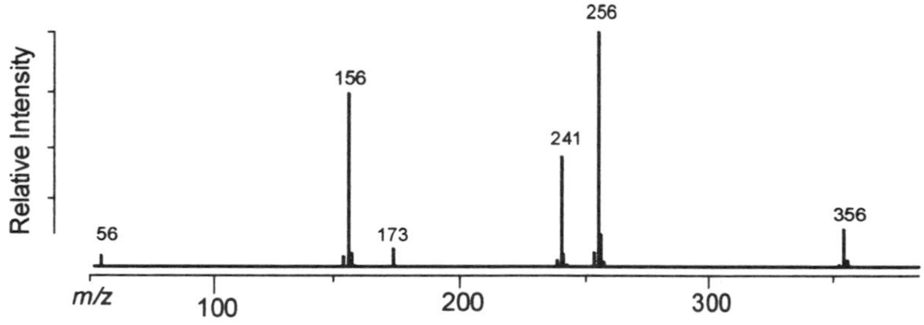

Fig. 5.14 The EI mass spectrum of $Fe(acac)_3$.

5.6 Identify the volatile liquid product from the reaction of tin tetrachloride with triethylaluminium based on the mass spectrum in Fig. 5.15.

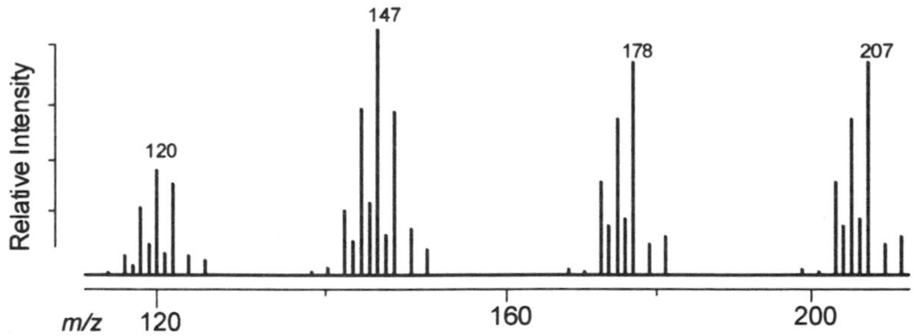

Fig. 5.15 The mass spectrum of a product from the reaction of $SnCl_4$ and $AlEt_3$.

Further reading

R. Adams, R. Gijbels and R. vonGrieken (eds.), Inorganic mass spectrometry, Wiley (1988).

M.R. Litzow and T.R. Spalding, Mass spectrometry of inorganic and organometallic compounds, Elsevier (1973).

6 Putting it all together

Introduction

It is rarely sensible, and frequently not possible, to rely on a single spectroscopic method to provide an unambiguous answer to the identity of an unknown product. Usually, the more techniques that are used the more confident you can be in your assignment of the identity of a compound and its potential structure. This short chapter provides some worked problems where a number of techniques are brought to bear on the same problem.

Problem 6.1

The reaction of $(Me_2HSi)_2S$ with methanol results in a gaseous product which we wish to identify. Firstly, we need to identify which techniques are going to furnish the most useful information. The presence of a variety of protons makes NMR spectroscopy a likely candidate, provided we can find a solvent in which this gas is soluble. For a solid sample we would try and obtain the molecular composition by elemental analysis, for a volatile sample mass spectrometry is more straightforward. Figure 6.1 shows the ^1H NMR and mass spectra of this material.

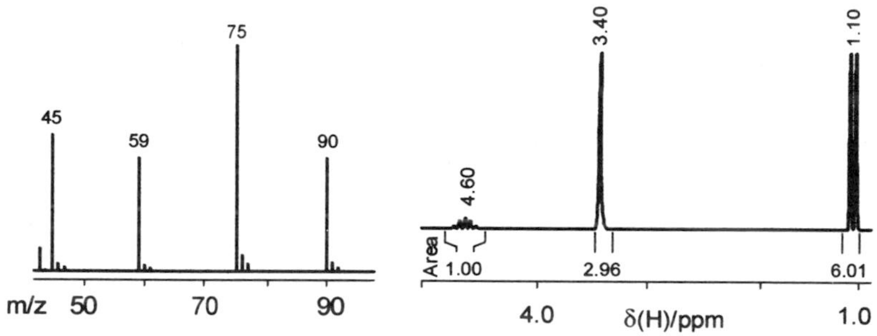

Fig. 6.1 The ^1H NMR and mass spectra of the product from the reaction of $(Me_2HSi)_2S$ with methanol.

There is no right or wrong way to analyse these problems, similarly there is no single 'entry-point' to the problem. So we will start with the proton NMR data and look for confirmation, or otherwise, that the Me_2HSi unit is still intact. We would predict that the proton NMR spectrum of this unit should consist of a signal for the methyl protons which will couple with the hydride to give a doublet. The data in Fig. 3.4 suggest that this peak should be between –1 and 1.2 ppm. The hydride signal could occur over a much wider chemical shift range (–2 to 4 ppm) but should be readily identified since it will appear as a septet due to coupling with the six methyl protons. These predictions are consistent with the peaks at *ca.* 1.1 and 4.6 ppm suggesting that this unit is still present in the product.

If we now turn to the mass spectrum we should see a peak due to this fragment at m/z = 59—which we do! The two peaks at higher m/z values appear at 75 and 90 amu, that is 16 and 31 mass units higher. Sixteen amu corresponds to the mass of oxygen, and 31 to methyl + oxygen. So we can explain the mass spectrometry data if we assume the product is $Me_2HSiOMe$. Now this explains the final peak in the proton NMR spectrum; the methoxy-protons will resonate at a higher frequency due to the presence of the more electronegative oxygen atom, and 3.4 ppm is fairly typical.

We can check that our formulation of the product is okay by carrying out some other form of spectroscopy and checking that this fits the data as well; if it doesn't then we need to re-think!

In this case we might choose to record the gas-phase IR spectrum, which results in bands around 3000(s), 2130(m), 1255(m), 1100(s), 910(s), and 770(m) cm^{-1}. Of these bands we only try to assign the characteristic rather than the fingerprint peaks. The bands around 3000 cm^{-1} must be due to ν(C–H) and these will be accompanied by δ(C–H) absorptions, which may explain the peaks at 1255 and 1100 cm^{-1}. According to the data in Fig. 2.20 the band at 2130 cm^{-1} must be due to ν(Si–H). There are no unexplained features and so we can be fairly confident in our assignment.

Problem 6.2

The photochemical reaction of triphenylphosphine with the dimeric complex $[(\eta^5\text{-}C_5H_5)(CO)_3Mo]_2$ in benzene results in a red complex, the spectroscopic data for which are summarised in Table 6.1. What is the resulting complex?

Method	Data obtained
Elemental analysis	62.5 % C, 4.2% H, 6.5% P
^{1}H NMR δ (ppm)	4.9 (m), 5.2 (s)
IR ν (cm^{-1})	1853 (s), 1830 (s)

Table 6.1 Elemental and selected spectroscopic data for Problem 6.2.

This time we will start with the elemental analysis figures. The empirical formula can be obtained by taking the percentage analysis figures and dividing each by the relative mass of the elements involved to give us the molecular ratio in the analysed compound.

$$\begin{aligned} &C:H:P \\ &\frac{62.5}{12}:\frac{4.2}{1}:\frac{6.5}{31} \\ &= 5.21:4.20:0.21 \\ &= 24.8:20:1 \end{aligned}$$

Thus the empirical formula of this compound appears to be $C_{25}H_{20}P_1Mo_xO_y$. The fact that there is one phosphorus atom in the formulation suggests that there is one triphenylphosphine ligand in the resulting complex. If this is so then we can account for $C_{18}H_{15}P$ which leaves $C_7H_5Mo_xO_y$. We now turn to the IR data which shows two bands typical of ν(C≡O) of a metal carbonyl. The presence of two carbonyl ligands would account for two more of the carbons in the empirical formula leaving C_5H_5 plus some molybdenum and, maybe, oxygen. The starting complex contained the cyclopentadienyl ligand and it would appear that the product does as well. So the product appears to be $[(\eta^5\text{-}C_5H_5)PPh_3Mo(CO)_2]$.

	Calculated	Found
C	62.6	62.5
H	4.2	4.2
P	6.5	6.5

Table 6.2 Calculated and observed elemental analysis for $[(\eta^5\text{-}C_5H_5)PPh_3Mo(CO)_2]$.

If we calculate the expected elemental analysis for this complex it agrees with that obtained experimentally, as shown in Table 6.2. This complex is consistent with the IR data which shows two ν(CO) bands and it is also consistent with the proton NMR data. The multiplet at 4.9 ppm, in the aromatic region, is due to the protons on the phenyl rings and the singlet at 5.2 ppm is due to the cyclopentadienyl ligand which is spinning so that only one peak is observed—everything appears to fit. But this is not the correct answer! An electron count of the postulated product gives

Mo(I)	d^5
Cp^-	6
2 × CO	2 × 2
PPh_3	2
Total	17 electrons

Most 17-electron organometallic species are unstable and dimerise, which is exactly what happens in this case. The dimer also fits the elemental analysis, IR and ^{1}H NMR data, so the identity of the product of photolysis is $[(\eta^5\text{-}C_5H_5)PPh_3Mo(CO)_2]_2$. Which acts as a timely reminder—however you interpret spectroscopic data the answer must make sense chemically!

Further problems

Element	% Composition
C	7.1
H	1.8
N	8.5
Cl	63.6
P	18.8

Table 6.3 Elemental analysis for the product of reaction between PCl_5 and CH_3NH_3Cl.

6.3 The molecular product from the reaction of PCl_5 with CH_3NH_3Cl analysed as shown in Table 6.3. The ^{35}Cl NQR spectrum exhibited two peaks in the ratio 2:1 and the ^{31}P NMR consisted of a singlet at −78 ppm. Solution-phase IR and Raman spectra were recorded and strong absorptions were observed at 2996, 2941, 1210, 1184, 1162, 847, 658, and 575 cm^{-1} and 3000, 544, and 458 cm^{-1} respectively. Assign the spectroscopic data and hence determine the identity of the product.

6.4 $Mo(CO)_6$ reacts with $CH_2{=}CHCH_2SCN$ in acetonitrile to yield a red solid which analyses to the empirical formula $MoC_{10}H_{11}O_2SN_3$. The Nujol mull IR and ^{1}H NMR spectra are shown in Fig. 6.2. What is the product?

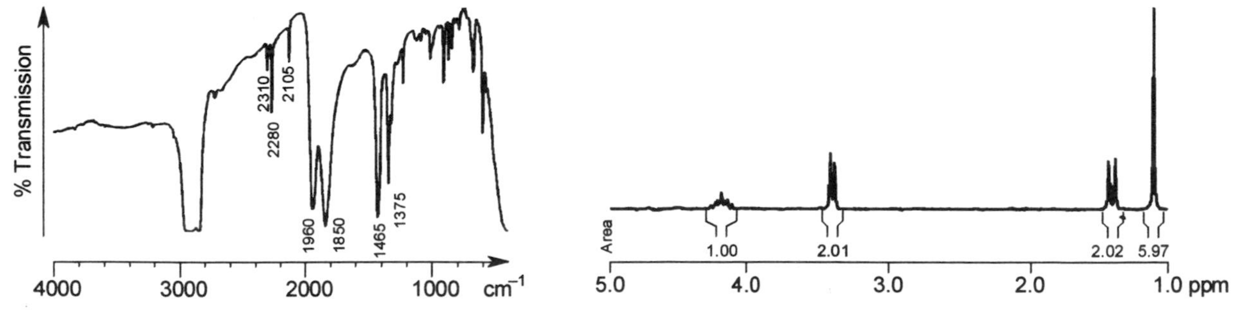

Fig. 6.2 The Nujol mull IR and ^{1}H NMR spectra for Question 6.4.

Answers to questions

1.1 $\lambda = c/\nu$ gives 3 m; $E = Lh\nu$ gives 0.04 J mol^{-1}, i.e. 4×10^{-5} kJ mol^{-1}.
1.2 2000 cm^{-1} = 5×10^{-4} cm = 5×10^{-6} m.
1.3 Changes in magnetic states are of lower energy than ESR transitions so we might expect to see these as fine structure in ESR studies.
1.4 $E = 4.42 \times 10^{-19}$ J so $N_{upper}/N_{lower} = \exp(4.42 \times 10^{-19}\text{ J}/1.381 \times 10^{-23}\text{ J K}^{-1} \times 298\text{ K}) = 107.4$.
1.5 The broadest bands will be seen for the quickest relaxation process which suggests UV–visible.

2.1 (i) BrCl—lighter constituent atoms; (ii) Br_2^+—higher bond order; (iii) NH_3—lighter constituent atoms.
2.2 (i) H_2S, 3 [$3N$–6]; (ii) CO, 1 [$3N$–5]; (iii) NH_3, 6; (iv) $Ni(CO)_4$, 21.
2.3 (i) and (ii) two each since both have two Pt–Cl bonds.
2.4 Water—all three should be IR active; CO_2—all but ν_1 are IR active.
2.5 ν(CO) for $Ni(CO)_4$ should be higher (actually 2058 cm^{-1}) than $Co(CO)_4^-$ (1890 cm^{-1}) due to greater backbonding in the cobalt carbonylate anion.
2.6 One; only the degenerate (T_{1u}) mode is IR active.
2.7 $A_{1g} + E_g + T_{1u}$—exactly the same as for the C≡O bonds. The symmetry operations will affect the M–C bonds in the same way as the C–O bonds since they are co-linear.
2.8 Considering M–Cl sections only and referring to Table 2.4 suggests (i) one and (ii) two.
2.9 Some of the NO_2 ligands are bound as NO_2^- (1347 and 1325 cm^{-1}) and some as –ONO (1387 and 1206 cm^{-1}).
2.10 From the entry in Table 2.4 for tetrahedral CCl_4 the Raman spectrum should consist of $A_1 + T_2$ stretches and T_2 bend. From the depolarisation ratios the bands at 218 and 314 cm^{-1} must be T_2 and at 459 cm^{-1} will be A_1.
2.11 KrF_2 must be linear, there is mutual exclusion of IR and Raman bands.
2.12 ν(CO) is higher in the benzene complex; therefore this is a less good π-donor ligand.
2.13 (i) Pd–Br vibrations will be at very low energies so use Raman; (ii) Raman (IR inactive); (iii) IR.
2.14 1083 and 1065 ν(P=O); 577, 561, and 551 ν(Sn–F); 472 and 448 ν(Sn–O); the complex is the *cis*-isomer.

3.1 4.2 – 4.1 ppm = 0.1 ppm × 100 MHz = 10 Hz.
3.2 (i) H_2O, one; (ii) $CH_3CH_2SCH_3$, three; (iii) $(CH_3)Pt(P(CH_3)_3)Cl_2$, two.
3.3 (i) two: CO *trans*-CO and CO *trans*-Br; (ii) three: C_2H_4, and CH_2 and CH_3 of PEt_3.
3.4 (i) 1.5 ppm; (ii) 0.2 ppm × 200 MHz = 40 Hz.
3.5 PH_3—pyramidal shape like NH_3; ^{1}H NMR, one proton chemical environment split by one phosphorus to give a doublet; ^{31}P NMR, one phosphorus environment split by three protons into a quartet.
3.6 AX—both doublets; AX_2 A spectrum is a triplet, X spectrum a doublet; AX_3 A spectrum is a quartet, X spectrum a doublet.
3.7 (i) AX_6; (ii) A_3B_2X; (iii) AA'XX'.
3.8 (i) $J_{PH} = 0$ gives a doublet of doublets; (ii) $J_{PH} = J_{PP}$ gives a triplet of doublets.

3.9 All four carbons are equivalent and would therefore give rise to a singlet. There will be no coupling to protons since the spectrum is proton decoupled. Coupling to ^{117}Sn and ^{119}Sn nuclei will occur to yield satellites and a spectrum similar in appearance to that for the ^{1}H NMR. Tin is less electronegative than silicon so we expect a chemical shift lower than for $SiMe_4$, i.e. below 0 ppm.

3.10 182 ppm terminal COs, 234 ppm bridging COs; weighted average $(182 \times 6 + 234 \times 2)/8 = 195$ ppm.

3.11 Si satellites are just visible on peak D, but not others, so it is near the Si nucleus.

3.12 Yes; *trans*-isomer has two equivalent Cl nuclei so one peak, *cis*-isomer has two non-equivalent nuclei so two peaks.

3.13 $E = h\nu = g\mu_B B$ so $g = (h\nu)/(\mu_B B) = 6.626 \times 10^{-34}$ J s $\times$ 9136 $\times 10^6$ s^{-1})/(9.274 $\times 10^{-24}$ J T^{-1} $\times$ 0.325 T) = 2.0084.

3.14 (i) I_2, no ESR spectrum; (ii) I_2^-, yes; (iii) NO, yes (17 valence electrons in total); (iv) VO^{2+}, yes (d^1 ion).

3.15 Interaction with the ^{14}N nucleus ($I = 1$) gives three lines of equal intensity, each are split into triplets by two equivalent ^{1}H nuclei ($I = ½$). Measured from the figure, $g = 2.0059$, aN=16 and aH=3.5 mT.

3.16 All signals split into doublets due to Rh coupling. ^{31}P NMR is not unequivocal, NQR would help.

3.17 Product is $(PPh_3)_2Pt(C_2H_4)$. Pt-coupling seen on π-bound ethylene signals which are singlets due to rotation.

4.1 $A = \varepsilon cl$ so $\varepsilon = A/cl = 1.5/(2 \times 10^{-3}$ mol dm^{-3} $\times$ 1 cm) = 750 mol dm^{-3} cm^{-1}.

4.2 Since Δ_t is smaller than Δ_o we should expect more tetrahedral complexes to be high spin than octahedral.

4.3 Three, (d_{xz}, d_{yz}) to d_{z^2}, d_{xy}, and $d_{x^2-y^2}$.

4.4 $[V(CN)_6]^{3-}$ is a d^2 complex so using the diagram in Fig. 4.17, $\Delta_o/B = 40$, $E/B = 37.3$, so Δ_o= 23600 cm^{-1} and $B = 590$ cm^{-1}.

4.5 $\beta = 590/855 = 0.69$. This suggests a reasonable degree of delocalisation of the electron density.

4.6 d–d transitions in a square-planar complex will be forbidden since $\Delta l \neq \pm 1$, and are Laporte forbidden since the molecule has a centre of symmetry. Individual transitions may also be spin forbidden.

4.7 $x = 2$, $y = 1$ gives the nearest estimate for the mixed complex.

4.8 The lowest CT band for $[ReO_4]^-$ will be lower since this may be reduced more easily than $[MnO_4]^-$.

4.9 The two lowest energy bands are d–d transitions, the third is either d–d or a forbidden CT band. The last three are CT; using the first of these and the optical electronegativity formula (eqn 4.4) gives χ_{opt}(Ru(VI)) = 2.8.

4.10 The bands at 630 and 440 nm are d–d, using the first one gives Δ_o *ca.* 15900 cm^{-1} which suggests the complex in solution is $[Cr(H_2O)_4Cl_2]^+$.

5.1 $^{35}Cl^{79}Br$ $m/z = 114$, relative intensity = 3; $^{35}Cl^{81}Br$ and $^{37}Cl^{79}Br$, 116, 4; $^{37}Cl^{81}Br$, 118, 1.

5.2 PMe_3^+, PMe_2^+, PMe^+, P^+, Me^+, CH_3^+, CH_2^+, CH^+, C^+, H^+.

5.3 $m/z = 64 = SO_2^+$, $48 = SO^+$, $32 = S^+$ or O_2^+, or SO_2^{2+}.

5.4 $B_2H_6^{2+}$ pattern would be the same as $B_2H_6^+$ but at half the m/z value and with half the spacings between all peaks.

5.5 $m/z = 56 = Fe^+$, 156 = $Fe(acac)^+$, 173 = $Fe(acac)OH^+$, 241 = $[Fe(acac)_2–CH_3]^+$, 256 = $Fe(acac)_2^+$, and 356 = $Fe(acac)_3^+$.

5.6 Most abundant isotope of tin is ^{120}Sn so m/z = 120 = Sn^+, 147 = $SnEt^+$, 178 = $SnEt_2^+$, and 207 = $SnEt_3^+$. No peak shows the presence of Cl so the product is most likely $SnEt_4$ and the parent ion is not seen.

6.3 Analysis gives an empirical formula of CH_3NPCl_3. NQR suggests two chloride environments in the ratio 2:1. ^1H NMR suggest all protons are equivalent, i.e. CH_3 group. IR/Raman have no coincidences, a centre of inversion is likely which is not possible for the monomer. Product is $[CH_3NPCl_2]_2$ containing a N–P–N–P ring with axial and equatorial chlorides on the phosphorus and methyl on nitrogen.

6.4 The IR data suggest the presence of at least two carbonyl ligands (1960, 1850 cm^{-1}) and two $\nu(C{\equiv}N)$ modes (2310 and 2280 cm^{-1}). The ^1H NMR spectrum exhibits a typical methyl group resonance at 1.1 ppm, so IR and NMR suggest two MeCN ligands (which from IR must be in a *cis*- not *trans*-arrangement). The NMR also shows signals due to a π-bound allyl unit. Removing the two CO, two MeCN, and one allyl ligand from the empirical formula leaves the SCN^- ligand, which accounts for the band in the IR at 2105 cm^{-1} [$\nu(-N{=}C)$] typical of nitrogen-bound SCN^-. The product is $[Mo(CO)_2(MeCN)_2(\eta^3\text{-allyl})(SCN)]$.

Index

Chemical formulae for compounds and ligands are listed separately on the next page.